Lecture Notes on Data Engineering and Communications Technologies

Volume 42

Series Editor

Fatos Xhafa, Technical University of Catalonia, Barcelona, Spain

The aim of the book series is to present cutting edge engineering approaches to data technologies and communications. It will publish latest advances on the engineering task of building and deploying distributed, scalable and reliable data infrastructures and communication systems.

The series will have a prominent applied focus on data technologies and communications with aim to promote the bridging from fundamental research on data science and networking to data engineering and communications that lead to industry products, business knowledge and standardisation.

** Indexing: The books of this series are submitted to ISI Proceedings, MetaPress, Springerlink and DBLP **

More information about this series at http://www.springer.com/series/15362

Dmytro Ageyev · Tamara Radivilova ·
Natalia Kryvinska

Editors

Data-Centric Business and Applications

Evolvements in Business Information Processing and Management (Volume 3)

 Springer

Editors
Dmytro Ageyev
Department of Infocommunication
Engineering, Faculty of Infocommunication
Kharkiv National University of Radio
Electronics
Kharkiv, Ukraine

Tamara Radivilova
Department of Infocommunication
Engineering, Faculty of Infocommunication
Kharkiv National University of Radio
Electronics
Kharkiv, Ukraine

Natalia Kryvinska
Department of e-Business, Faculty
of Business, Economics and Statistics
University of Vienna
Vienna, Austria

ISSN 2367-4512   ISSN 2367-4520   (electronic)
Lecture Notes on Data Engineering and Communications Technologies
ISBN 978-3-030-35648-4   ISBN 978-3-030-35649-1   (eBook)
https://doi.org/10.1007/978-3-030-35649-1

This Springer imprint is published by the registered company Springer Nature Switzerland AG
The registered company address is: Gewerbestrasse 11, 6330 Cham, Switzerland

Preface

Modern methods of business information processing with using the methods of artificial intelligence, big data, their cloud-based storage and processing open new opportunities for doing business based on information technologies. In the third volume, we continue to study various aspects of it—technological as well as business and social.

The first chapter "FinTech, RegTech and Traditional Financial Intermediation: Trends and Threats for Financial Stability" deals with the issues of main directions, challenges and threats of development of the newest financial technologies—FinTech, RegTech and traditional financial intermediation. The authors analyze the modern ecosystem and Fintech's global trends, their influence on the transformation of principles, models and emergence of new threats of traditional financial intermediation. It is important to adapt the regulatory system to ensure a balance between the introduction of financial innovation, economic growth and financial stability. It is reasonable to use the services of RegTech companies.

In the work called "Data-Centric Formation of Marketing Logistic Business Model of Vegetable Market Due to Zonal Specialization," the authors concern the scope of methodical material for the formation of a new business model of the vegetable market of Ukraine. The chapter defines the option of transporting products, finding the optimal supply of products from the manufacturer to the consumer. The analysis of trends in the production and consumption of vegetable products in the regions of Ukraine based on open data sources. To solve the center location optimization problem, according to the proposed criterion, the center of gravity method was used. As a result, the geographical coordinates of the warehouses are calculated, in which the total annual transportation costs of the cargo in the system under consideration will be minimal.

The chapter authored by Kovalskyi et al. "Optimization of the Process of Determining the Effectiveness of Advertising Communication" presents the description of advertising perception process by a potential consumer. The authors proposed a method for choosing weight factors, which is based on the basis of pair comparisons using the method of numerical consistency and the matrix of pair comparisons. The weighting of factors was optimized, and an optimized model of the priority influence of factors influencing the effectiveness of advertising communication was synthesized. The alternative variants of the process of determining the effectiveness of advertising communication are designed using the method of multi-criteria optimization. The criterion for choosing the optimal alternative is the maximum value of the unified functional. As a result, the conducted studies provide an opportunity to optimize and calculate alternatives in other processes.

In the next chapter titled "Using Recurrent Neural Networks for Data-Centric Business," the authors review and propose the Recurrent Neural Networks in data-centric business. The results of research and their analysis show that in the future for data-centric business as a model (building tool), it is better to use Recurrent Neural Networks, exactly Long Short-Term Memory network architecture. The architecture of Gated Recurrent Unit has proved to be a good alternative, but due to its innovation in some cases, it is inferior in quality of work. In case of using Bidirectional Recurrent Neural Networks architecture, special attention should be paid to optimizing the process of network training (synthesis). One of the solutions, for example, can be the use of neuroevolutionary approaches that have proven themselves well when working with Recurrent Neural Networks. Such approaches will help to significantly reduce the training time of the network while maintaining the accuracy of forecasting.

The chapter "System Model for Decision Making Support in Logistics and Quality Management Business Processes of Manufacture Enterprise" focuses on minimizing the time and cost for decision-making support in the enterprise's logistics business processes, taking into account quality management methods. To obtain the results, systems analysis methods were used to develop a procurement management model, language algebra and algebra of algorithms for describing the sequence of solving local procurement management problems and methods of a full-factor experiment for assessing the influence of the environment on the components of the logistics chain. The work presents the business processes in the management of a manufacturing enterprise with regard to quality control. A system model of decision support is proposed, including logistic processes and quality management tasks. These components of the model are signed by the algebra of algorithms of equations, taking into account business processes as stations and processes as a result of the transfer condition.

The work entitled "Machine Learning Methods Applications for Estimating Unevenness Level of Regional Development" deals with the issues of socio-economic development of the region. Special attention is paid to the assessment and analysis of the level of unevenness at different levels of the hierarchy. It is proved that the relatively low position of the country is determined by many different external and internal factors, but one of the most significant is the uneven development of the region. The main goal of the research is to build a set of models for assessing and forecasting the level of unevenness in the socio-economic development of regions. The complex is based on the methods of modeling multi-dimensional objects. The proposed models should be introduced into the decision-making process and become the basis for an effective regional socio-economic policy.

The work authored by Cherednichenko et al., "Formal Modeling of Decision-Making Processes Under Transboundary Emergency Conditions," is dedicated to the problems of modeling of decision-making processes in the case of emergency occurrence. The transboundary nature of emergencies implies that data sources for environmental monitoring may not be available, and data collected from available sources may be incomplete, inaccurate or duplicate. This study presents three types of problems related to emergency management: prevention of emergency situations; immediate response to the occurrence that just happened; and elimination of accidental consequences. A formal presentation of these problems is the basis for developing a decision support system. The chapter presents the rules that allow to transform the conditions of a formal statement of a decision task into knowledge used by the environmental monitoring system.

The next chapter "Decision-Making Support System for Experts of Penal Law" substantiates the relevance of the task of developing an expert decision-making support system for criminal law specialists. For the effective implementation of the expert system, a special data set structure was proposed, based on the basic properties of the Criminal Code Article. Organized data sets allow you to use them in the future for data mining. The decision tree method was chosen as a method for extracting new knowledge from fact bases. Two main ways of its construction are considered—manual and machine learning. The developed system is based on a client–server architecture. The main modes of operation of the dialog part of the system, obtaining solutions and interpreting the results obtained are described. The final part contains a discussion of the possibilities and limitations of the developed system, the positive and negative aspects of the methods and technologies used.

The research done in the chapter "Infocommunication Tasks of University Information System" is aimed at analyzing the modern state and problems of information communication technologies in higher educational institutions. So, it should be noted that the real rates of change in the global educational sector (especially in its private part) are much faster than stereotypes. Thus, higher educational institutions try to meet these changes early, although in most cases the infrastructure and information systems of higher educational institutions cannot do this. Consequently, there is an urgent scientific task to redefine the needs of infocommunications in higher education institutions using their information flows in order to determine the actual modern requirements for information systems and the infocommunication infrastructure of higher educational institutions.

And, the final chapter "Advanced Methods and Means of Authentication of Devices for Processing Business Information" is devoted to the analysis of the current state of methods and means of authentication of electronic devices for tasks of processing business information. Transmission, transformation and storage of business information are carried out by a wide range and a growing number of devices both in global networks and beyond. This requires adequate mutual authentication between devices to enhance cybersecurity, as well as reliable authentication of various electronic devices for digital media expertise, detection of counterfeit products, etc. The general approach to authenticating electronic devices based on their individual differences allows us to solve various business informatics applications. Tasks that are solved on the basis of authentication are proposed to be divided into six classes, and electronic devices based on their individual differences are used as a general approach. Besides, the production of electronic devices ensures their operation within fault tolerance and allows for physical non-identity.

Dmytro Ageyev
Department of Infocommunication Engineering
Faculty of Infocommunication
Kharkiv National University of Radio Electronics
Kharkiv, Ukraine
e-mail: dmytro.aheiev@nure.ua

Tamara Radivilova
Department of Infocommunication Engineering
Faculty of Infocommunication
Kharkiv National University of Radio Electronics
Kharkiv, Ukraine
e-mail: tamara.radivilova@nure.ua

Natalia Kryvinska
Department of e-Business, Faculty of Business
Economics and Statistics
University of Vienna
Vienna, Austria

Department of Information Systems
Faculty of Management
Comenius University in Bratislava
Bratislava, Slovakia
e-mail: natalia.kryvinska@univie.ac.at

Contents

FinTech, RegTech and Traditional Financial Intermediation: Trends and Threats for Financial Stability

Natalia Pantielieieva⬤, **Myroslava Khutorna**⬤, **Olesia Lytvynenko**⬤ **and Liudmyla Potapenko**⬤

Abstract The paper deals with the issues of main directions, challenges and threats of development of the newest financial technologies—FinTech, RegTech and traditional financial intermediation. The authors analyze modern ecosystem and Fintech's global trends, their influence on the transformation of principles, models and emergence of new threats of traditional financial intermediation. The trends of FinTech distribution in Ukraine are considered. The areas of activity of traditional financial intermediaries, which require the use of the latest technological tools, the possible effects of the influence of modern regulatory standards on the banking sphere are studied. The authors describe features of RegTech application for improvement financial intermediaries risk management systems, usage of innovative regulatory technology tools that can enhance the quality and efficiency of financial institutions compliance systems as a prerequisite for increasing the protection of depositors, creditors and investors interests. New opportunities for cooperation between traditional financial intermediaries and RegTech companies are examined. Post-crisis world-view changes in the formulation of targets for the use of financial innovations and FinTech are substantiated. The spectrum of potential financial stability risks posed by FinTech, which requires systematic monitoring by entities providing financial stability, is determined.

Keywords Digital divide · Digital economy · Digital transformations · Financial innovation · Technological innovation · FinTech · RegTech · Financial regulation and supervision · Financial stability · Financial services

N. Pantielieieva (✉) · M. Khutorna · O. Lytvynenko · L. Potapenko
Cherkasy Institute, Banking University, Cherkasy, Ukraine
e-mail: nnpanteleeva2017@gmail.com

M. Khutorna
e-mail: lmiroslava7@gmail.com

O. Lytvynenko
e-mail: olesia.lytvynenko@gmail.com

L. Potapenko
e-mail: milavit7@ukr.net

© Springer Nature Switzerland AG 2020
D. Ageyev et al. (eds.), *Data-Centric Business and Applications*,
Lecture Notes on Data Engineering and Communications Technologies 42,
https://doi.org/10.1007/978-3-030-35649-1_1

1

1 Introduction

The emergence and rapid expansion of financial technologies (FinTech) are caused:

On the one hand by the existence of unresolved problems that have already become chronic in nature—financial crises with a tendency to decrease cycle times and financial instability, high cost of financial resources, limited access to financial services, low efficiency of traditional financial model mediation, poor transparency of national economies and the growth of their shadow segments, corruption, etc.;

And on the other hand by the formation of favorable conditions at the expense of stable growth dynamics of computing, resource, analytical possibilities of information and communication means and technologies, their perception and support by a new generation of digital society.

The described processes cause the active formation and effective use of FinTech and RegTech for deepening the processes of digital transformation of traditional financial intermediation and do not break financial stability and security.

The theoretical and methodological basis for studying the problems of FinTech distribution is formed by papers of T. Philippon, S. Darolles, S. Bunea, B. Kogan, P. Furche, P. T. Harker, S. Creehan, Y. Ma, M. Demertzis, D. He, J. J. Cortina and others. It is important to note that the development of FinTech is a complex issue that is a general trend in the service economy and touches on not only financial intermediation. Thus, the peculiarities of the formation of a service-dominant economy are considered in the papers of Kaczor and Kryvinska [1]. The specifics of FinTech implementation in the activity of enterprises in the real sector are described in detail in the writings of Kryvinska and Gregus [2–4]. The methodological principles for establishing effective forms of interaction between clients and service providers on the basis of the introduction of technological innovations are given in the papers of Molnar and Molnar [5]. Particular attention deserves the research devoted to the search for the most effective applied methods to evaluate and optimize the effectiveness of innovations introduction into activities of economic agents, including financial ones. The development of such scientific and methodological approaches can be found in the papers of Kirichenko et al. [6, 7], Dobrynin et al. [8], Bulakh [9], Tkachenko [10], Ageyev [11], Al-Ansari and Qasim [12], Wehbe [13], Al-Dulaimi and Al-Dulaimi [14].

Regarding the development of FinTech, according to Schindler [15], the development of financial intermediation industry is constantly accompanied by financial innovations. The problems of digital adaptation of banking and financial sectors are examined by Darolles [16], Bunea et al. [17]. Cortina and Schmukler [18] highly evaluate the benefits of FinTech and emphasize the importance of balanced regulation to ensure financial stability. Treleaven [19] notes that the proliferation of FinTech has stimulated the emergence of a new generation of RegTech companies that provide software to regulate financial companies. Regional aspects of FinTech expansion and state policy regarding it, in particular in China, are studied by Tsai [20], in Asian countries—by Creehan and Borst [21], in the European Union—by Demertzis et al. [22]. The issues of transformation of business models and formation of ecosystems, identification of potential risks of financial intermediation, monitoring and adaptation of the regulatory system for ensuring financial stability and security are of the same importance as the studying the trends of FinTech distribution.

2 Proposed Methodology

The methodology of the research is based on the dialectical method of cognition, scientific generalization, institutional, comparative, analytical and systematic approaches to the study of the processes of financial services availability development and increase, formation of competition-oriented and openness of financial markets due to FinTech potential, adaptation of the regulatory system and modernization of the legal framework.

The paper aims to analyze development and to identify global trends of financial technologies and their impact on the principles and institutional, functional and social transformations of financial intermediation, to ground main issues of regulation monitoring and adaptation to ensure financial stability and security.

3 FinTech—A Driver of Financial Intermediation Evolution

3.1 Cosystem and Global Trends of FinTech

FinTech is technologically enabled financial innovation that could result in new business models, applications, processes, products, or services with an associated material effect on financial markets and institutions and the provision of financial services [15]. FinTech current ecosystem is formed by FinTech-startups, technology developers, government, financial institutions, consumers of financial products and services. Developed FinTech ecosystem promotes financial innovations, cooperation and competition, stimulates economy and makes financial services more accessible. FinTech ecosystem is determined by: (1) infrastructure; (2) favorable legal and regulatory support; (3) access to capital and investments; (4) human resources for development and maintenance. In particular, infrastructure is characterized by the development level of information and communication technologies, access to the Internet and mobile communications, development of financial intermediation payment and agency networks, implementation of AML, KYC and FATF digital identification tools. These factors determine the country's readiness for FinTech development. The position of financial institutions (banks, insurance companies, stock exchanges, venture funds) is determined by their innovative strategy and the decision to establish partnerships with FinTech companies or to start a competitive struggle.

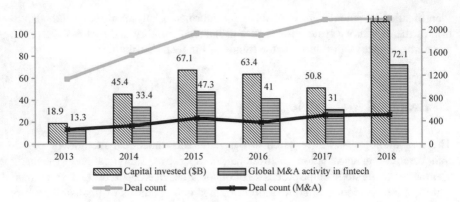

Fig. 1 Dynamics of global M&A activity in FinTech ($B) and the number of deals, 2012–2018. *Source* KPMG [23]

FinTech needs a lot of investment to accelerate the expansion. In 2018, the total investment in FinTech was $111.8 billion versus $18.9 billion in 2013. Shares increased by 15% compared to the same period of the last year, while financing increased by 220%. In 2018, consolidation trend was confirmed by the growth in Global M&A activity in FinTech—$72.1 billion (Fig. 1) and Global venture capital activity in FinTech with corporate participation—$23.1 billion. Global private investment (VC, PE and M&A) in RegTech in 2018 was $3.7 billion versus $1.1 billion in 2013 [23].

It should be noted that nearly 67% of FinTech's investments are formed on syndication bases. In addition, new startups have appeared in all areas of FinTech industry, including lending, banking and equity management. In particular, 16 new FinTech companies (neobanks Revolut, Monzo, Nubank, Chime, insurance company Policybazaar, investment company Tiger Brokers, Fintech-start-ups Plaid and Brex) joined "Unicorns", so it means they cost more than $1 billion [21]. The biggest investment was received by the Ant Financial payment company, which attracted $14 billion in 2018, accounting for 35% of Fintech's total funding in the last year [24].

Compared to 2014, the regional forces alignment has changed (Fig. 2). The most attractive region and the leader in FinTech investment is Asia, where the volume of venture agreements increased by 38% in 2018, amounting to $22.65 billion versus $12.4 billion and $3.53 billion in the United States and Europe respectively. According to the coverage of the population, Fintech services are chaired by China (69%), India (52%), Great Britain (42%), Brazil (40%) and Australia (37%), the United States (33%) [20].

An important factor in the development of FinTech ecosystem is a state continuous regulatory support, in particular liberalization of regulation and the creation of favorable conditions for business, access to markets and capital. The importance of this factor, as well as the need for compliance control and risk management, led to the development of RegTech projects as a separate sector of FinTech. During 2014—2018, global investments in RegTech increased almost fivefold. Increasing of CAGR

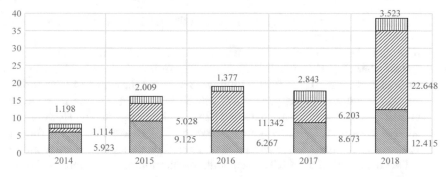

Fig. 2 Global VC-backed fintech funding by continent, 2014–2018, $B. *Source* FinTech Global [25]

by 48.5% from the point of view of invested capital has led to an increase in funding from $0.923 billion in 2014 to $4.485 billion in 2018 [26]. Investors turned to the old markets of North America and Europe, retreating from developing countries (Fig. 3).

Today, FinTech companies offer new business models using such technologies as Application Programming Interface (API), Artificial intelligence (AI), Machine learning, BigData, Internet of Things (IoT), Cloud Technology, Blockchain, Marketplace, Platform as a Service (PaaS), Software as a Service (SaaS). Google Analytics reports that over the past five years in Top 3 countries that are most interested in FinTech are: BigData—Spain, Morocco, Peru; Internet of Things (IoT)—Sweden, Ireland, Philippines; Artificial Intelligence (AI)—Pakistan, Bangladesh, Kenya; Blockchain—Ukraine, Vietnam, Russia; Application Programming Interface (API)—St. Helen Island, Nigeria, China. Singapore and Luxembourg are leaders in FinTech, RegTech and Insurtech.

The global FinTech ecosystem has been actively developing during 2018, its global trends were [27]:

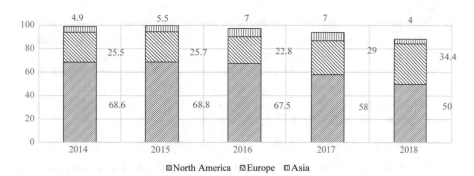

Fig. 3 Global RegTech deals by region, 2014–2018 (as a % of total number of deals. *Source* FinTech Global [25, 26]

(1) *Expansion of cryptocurrency* in the official financial sector and the emergence of new banking services;

(2) *ICO (Initial Coin Offering)*—increase in competition for investment resources and regulatory pressure will lead to the improvement of this mechanism. In particular, during 2017, $6.89 billion were invested through 514 ICO and the average amount of funds per project was $13.4 million. During four months of 2018—$7.57 billion, 301 ICO and $25.1 million respectively. However, the attitude towards ICO is ambiguous, since in 2017 almost 11% of the funding volume and 80% of ICO were fraudulent. Therefore, the regulators will continue to take measures to limit ICO along with the development of standards for this mechanism;

(3) *RegTech*—continuation of development, taking into account increased regulatory pressure in the financial sector, in particular the implementation of Basel III initiatives. RegTech scope will be in compliance, identity management and control, risk management, regulatory reporting, transaction monitoring, trading in financial markets;

(4) *SupTech (Supervisory Technology or RegTech for supervisors)* to provide support for the modernization of financial supervision, based on data with project implementation of data-input approach, real-time access, reporting utilities, gathering intelligence from unstructured data, regulatory submissions and data quality management [28];

(5) *Blockchain*—leads to the growth of applied decisions in many spheres of economy;

(6) *API*—compliance with the EU payment directive PSD2 (Remote Client Identification), Open API Specification for banks to open technology assets to the third parties, providing software integration and increasing their own presence in the digital environment;

(7) *Deepening partnerships between FinTech companies and banks* with further strengthening of the latest leading positions. It is interesting that during 2010–2015, the share of joint investments increased in North America (from 40 to 60%), APAC (from 7 to 16%), but in Europe it significantly decreased (from 38 to 14%) [29].

(8) *Intensification of the struggle between the world's leading FinTech leaders* (China, the USA, Germany and the United Kingdom). For example, according to the EY FinTech Adoption Index in 2017 in China the overall acceptance of FinTech was 69% of the population, awareness—98% digital-active consumers, concentration—39% of consumers had more than five FinTech services, while the USA had 33%, 74% and 15%, Germany—35%, 64% and 52%, the United Kingdom—42%, 42% and 43%, respectively.

It is expected that in 2019 the main tendencies, along with existing ones, will be:

(1) The growth of consolidation not only in payments and credit area, but also in Blockchain, in particular, the growth of investments in the development of specific decisions on their basis, it allows to use the scale effect and expand the international presence;

(2) Investors will continue to invest in FinTech, but to minimize risks they will do it at the final stages of development and in start-ups with tested reputations;

(3) The growth of investments in Insurtech is expected, in particular in Asian countries, which are selected by leading insurance companies of the USA and Europe as pilot sites;

(4) Keeping the policy of openness of banking services will have a positive impact on the development of FinTech, on cooperation between banks and high-tech companies, which will lead to the implementation of new, more flexible and autonomous solutions for digital banking [23].

In general, all FinTech sector could be divided into two groups: (1) companies that offer financial services and compete with traditional intermediators; (2) companies that develop technology-based infrastructure and services. If the first group of FinTech companies creates an alternative medium to traditional financial intermediation, the second group of companies aims at the operating activities of the business (including the field of financial services) and its improvement through the application of innovative technological solutions. That is, on the one hand, FinTech companies form competitive pressure on the financial intermediation market, forcing traditional financial institutions to change their own business models. On the other hand, FinTech firms that develop technology-driven infrastructure and services can balance the competitive opportunities of traditional and high-tech financial intermediation. In the collaboration of traditional intermediaries and infrastructure FinTech companies, we see the potential for further development of the traditional financial intermediation sector.

3.2 Financial Intermediation of the Future: Principles, Models and Risks

Financial intermediation is the main institution of economic growth of the national economies of the world. At present, we can observe the violation of the constancy of financial intermediation structure, the basis of which for a long time form insurance companies, commercial banks, investment and pension funds. Financial intermediation increasingly goes beyond traditional financial institutions. The latter are competing with powerful IT companies (Amazon, Google, Apple, Alibaba, Microsoft, Facebook, etc.), manufacturing corporations (e.g. Samsung), trading networks (Starbucks or Walmart), and FinTech start-ups. Taking into account the speed of changing the financial intermediation format under the influence of FinTech, we can identify the following driving principles:

(1) Innovative modernization as a method of overcoming the negative effects of crisis phenomena, intensifying competition and disintermediation in the financial market;

(2) Strategic priority—the relevance of the introduction of digital technologies to the strategic priorities of financial intermediaries as institutes of development;

(3) Strategic partnership—digitalization in the financial sector aims at deepening of financial integration on the basis of public–private partnership;
(4) Transparency policy and reliability of digital technologies;
(5) Personalization—the study of value benchmarks and individual demand, the provision of personalized needs with attention to the behavioral aspects of consumers in financial intermediation;
(6) Fair, reliable regulation—a moderate approach to legal regulation, taking into account situational uncertainty, and self-regulation;
(7) Optimality of interests and risks balance—the desire and ability to get the maximum positive effect of financial intermediation under the conditions of permissible risks of potentially negative effects using digital technologies.

Significant changes in the financial intermediation ecosystem benefit to FinTech, as they provide strategic priorities for the new vector, change traditional operating models and increase competitive dynamics. FinTech is spreading in such segments of financial services as deposits and lending, insurance, payments, investment management, capital markets, market infrastructure. The financial upheavals caused by crises require a renewal and a search for a new regulatory paradigm that should focus on achieving financial stability, securing the financial security of the banking sector and overcoming the pro-cyclicality. The traditional areas of banking regulation are the implementation of international banking supervision standards recommended by the Basel Committee and their adaptation to national legislation. Today conditions we can state that the banking sector is developing under conditions of rather strict regulation and supervisory practice. Basel's requirements, as recommendations at the initial stage, gradually become obligatory. There is a gradual shift from a relatively limited regulatory model to the one, the subject of which is a wide range of financial intermediation risks, and ultimately to a system of macro-regulatory regulation and supervision based on a functional-oriented approach.

A traditional approach to overcoming and preventing risks, regardless of their origin, may have different-directional synergies, primarily for the banking sector (Table 1).

In general, review and regulatory complications have led to a significant increase of business spending in the field of traditional financial intermediation. At the same time, there is a steady decrease in the interest margin of financial institutions, in particular banks, which is the main source of their profits. This is especially true for banks in the Euro area, the interest margin of Ukrainian banks is at least 4% (Figs. 4, 5).

Thus, the strengthening of the regulatory burden (in particular, the implementation of Basel III, IFRS 9) on the activities of banks and financial institutions is a heavy burden for traditional financial intermediation.

Table 1 Expected impact of modern regulatory standards on the banking sector

Field	New demands	Possible negative consequences
Capital	Strengthening the capital base through – increasing capital minimum level – changes in the structure of capital	Increasing of the cost for maintaining capital and increasing of reserves Increasing the role of the regulator in the area of procyclic regulation
	Introduction of capital buffers (buffer of conservation and countercyclical buffer)	Increasing the risk of a capital reduction due to deductions and the gradual exclusion of subordinated debt from one's own capital
	Control of debt load by introducing the leverage ratio	Potential rise in the sources of capital formation
	Introduction of additional capital buffers for system-forming banks (system risk buffer and system importance buffer)	The necessity to reduce the volume of transactions with risky assets for banks that are not able to increase the level 1 capital Lowering of credit volume due to inability to replenish capital
Iquidity	Control over liquidity shocks absorption by introducing liquidity ratios of LCR and NSFR	Increasing of additional expenses for assets and liabilities balance The necessity for financing on the account of more stable sources on an on-going basis, that results in a decrease in the interest margin
	Conducting stress testing of liquidity risk	Increasing the share of liquid assets with a lower level of profitability Reducing interest rate risk will reduce the interest income, potential profit and profitability of the bank
Corporative management	Verification of the professional suitability of the council and board of a bank	Increasing the compliance costs due to its enhanced content
	Involvement of independent directors	
	Building an effective rewards system	
Disclosure of information	Implementation of IFRS 9	Increasing of operating costs
	Ensuring market discipline	

Source Developed by the authors on the basis of [30–32]

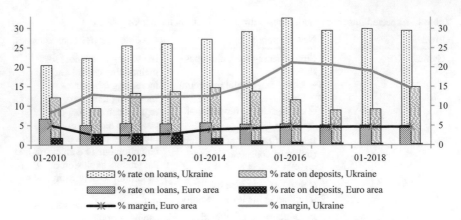

Fig. 4 The dynamics of interest rates on loans and deposits to individuals, as well as the interest margin of Euro-zone and Ukrainian banks, 2010–2019. *Source* National Bank of Ukraine, ECB statistical data. *Note* In calculations, loans and deposits of banks up to 1 year were taken into account

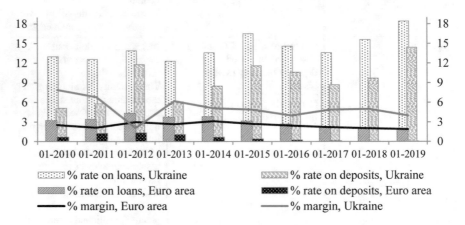

Fig. 5 The dynamics of interest rates on loans and deposits to legal entities, as well as the interest margin of Euro-zone and Ukrainian banks, 2010–2019. *Source* National Bank of Ukraine, ECB statistical data. *Note* In calculations, loans and deposits of banks up to 1 year were taken into account

3.3   FinTech in Ukraine: Beginning

Ukraine does not stand aside the global processes of financial intermediation digitalization. In 2018, there were more than 80 companies developing innovative financial solutions in Ukraine, with 58% of FinTech projects being established over the last three years. For the last three years the potential opportunities of FinTech have been implementing in the following areas: payments and money orders (31.6%); technologies and infrastructure (19.3%); lending (14%); platforms (7%); InsurTech (5.3%);

virtual and neobanks (5.3%); investment management (5.3%); mobile wallets (5.3%); Blockchain (3.4%); cryptocurrency (1.75%); RegTech (1.75%).

According to the Citi-Imperial Digital Money Readiness Index, in 2017 Ukraine ranks 64th. This index is based on 15 indicators belonging to different groups–institutional environment, enabling infrastructure, solution provision and propensity to adopt. In particular, Ukraine has shown steady high figures for the first and the last group, 73 and 59. The country has stepped up its efforts to create a technological and financial infrastructure that underpins the deployment and operation of digital money (70 versus 64 in 2016). However, solution provisioning (e-Government, e-Payments, e-commerce, Transit and Toll ways) has somewhat deteriorated (44 against 57 in 2016). According to the general rating, Ukraine is classified as "emerging" country, where there are basic regulation and infrastructure, but economy is unstable [33].

Among the factors, contributing to the development of FinTech in Ukraine, we can distinguish, on the one hand, the reduction of foreign banks presence, the cleaning and consolidation of the banking system and the loss of confidence in it, the reduction of the number of banks branches, and thus the deterioration of financial services availability. And on the other hand—highly intellectual human capital, adoption and support of global trends in the digitalization of financial intermediation, development of small and medium-sized businesses (funding of almost 45% of FinTech start-ups is made at the expense of own resources, but the need for investments is estimated at $40–70 million), legislative liberalization and regulator support (the NBU initiatives "Comprehensive Program of Ukraine Financial Sector Development Until 2020", "Promotion of FinTech in Ukraine", "BankID", "MobileID", etc.). In particular, the implementation of the EU directive of payment systems PSD2 provides support for payments security, protection of financial data and consumer rights, compliance with the uniform requirements and rules, transparent conditions for their functioning. At the same time, the NBU has the tasks to create a public register of authorized payment processing organizations, to develop norms and technical standards, etc.

Currently, the partners of FinTech-companies are domestic banks, in particular TASKOMBANK, Oschadbank, Alfa-Bank, Raiffeisen Bank Aval, FUIB, the sphere of their common interests is in the B2B segment, accounting for 59% of all payments, in particular for corporate clients—62.5%, small and medium-sized businesses—37,5%. At the same time, B2C segment—22.5%, B2B and B2C integrated—21.5%. The most striking examples in the Ukrainian FinTech field by GR Capital is the use of Apple Pay and Google Pay. Privatbank and Oschadbank were the first to activate these functions. Recently, monobank, which is considered an advanced financial start-up in Ukraine has started to use these functions. In 2016, Privatbank launched "KUB" service in the format of P2B-lending, and within two years, the amount of entrepreneurship financing amounted to 1 billion hryvnias. Accelerating of FinTech distribution requires a technological partnership. An illustration of this is the project "Open Banking Lab", with the participation of the NBU, OTP Bank and a non-profit incubator, which provides access to open (non-confidential) banking data and their API (Application Program Interface), which significantly reduces the cost of resources and time to innovate.

As it was already noted, investments are important for FinTech's development in Ukraine. In 2017, the focus of investments shifted from quantity to quality and the total volume of venture investments reached its maximum and amounted to $258.6 million, three times more than in 2016 ($88 million), with an average contract size of $3.3 million. In addition, 54% of investments were involved at the growth phase, and 15.4% were at an early stage of financing. By the number of deals, the largest share of high-value investments came to software ($119.7 million, 8 transactions) and online services ($13 million, 6 transactions). In 2016, the largest volumes of investments by the number of transactions were directed to online services ($29.5 million, 12 transactions) and mobile technology ($4.2 million, 4 transactions). There is a tendency of growing private equity investments, in particular, in 2017, 14 deals were made for $126.7 million.

Investing in Ukrainian start-ups through ICO's mechanism is a new trend for Ukraine. In 2017, $160.3 million were invested in the development of 19 start-ups: Neuromation—$70 million (AI), Rentberry—$30 million (E-Commerce). Moreover, venture funds are concentrated exclusively in the emerging crypto-business, establishing stronger links between private capital and industries based on Blockchain. The main ISOs are: e-commerce (6 transactions), online services (6 transactions), software (4 agreements) and others [34].

Recently, Blockchain Research Institute has recognized Ukraine as one of the 14 centers for the development of Blockchain innovations. In 2016, Blockchain auction "E-Auction 3.0" was the first Blockchain technology project used to reform the system of privatization and lease of state property in Ukraine. It gave the opportunity to buy state property with cryptocurrency. However, it should be noted that today the status of cryptocurrency in Ukraine is not defined. The NBU does not recognize cryptocurrency as means of payment and expects the adoption of the EU directives for their further harmonization in Ukraine. Nevertheless, in 2018, a progressive IT society initiated efforts consolidation to develop "Ukrainian Blockchain Policy Paper" to regulate cryptocurrency market. In July 2018, the Financial Stability Board endorsed the draft of the Concept of State Regulation of Crypto Foreign Exchange Operations in Ukraine, which defined the main aspects of legislative regulation in this field.

However, today the FinTech industry in Ukraine is regulated by traditional banking and/or financial legislation, which is objectively rather outdated, not flexible enough and does not meet the exact market needs. Most of FinTech sectors remain under-regulated or prohibited at all. This situation requires legal solutions and implementation of innovative products and services.

So, FinTech in Ukraine develops more on the principle "in contradiction of" than "due to", because the state understands the importance and irreversibility of FinTech initiatives implementation but provides very poor support and has a waiting attitude in this process. It means that the development direction and the goal are determined, but the movement of its main participants goes along parallel routes and at different speeds. Therefore, the FinTech initiatives in Ukraine have not yet become stable and systemic.

4 RegTech in Financial Intermediation: A Bonus or the Only Way for Its Further Development

The global financial crisis of 2008 has become a kind of litmus test for all faults both of management system and of regulation and supervision system of traditional financial mediators' activity. During the next 10 years, regulatory requirements were deepened and raised, which, in turn, led to revision of business internal standards. These increasing compliance-related cost pressures, coupled with low interest rates, a slow economy, and stagnant revenue growth, has resulted in significant downward pressure on returns [35].

In accordance with Chris Scinner's statements "a global bank has to deal with 185 regulatory changes per day, or a change every twelve minutes, and that a typical bank faces 128,000 regulations compared with technology firms that deal with just 27,000" [36]. Moreover, between 2008 and 2018, there has been a 1470% increase in regulatory changes in developed markets, which has led to 25–30% of the total workforce of financial institutions working under compliance functions.

Modern systems of internal control in banks and in other financial intermediaries are formed on the basis of the principles of continuality (is not limited by top management, but provides control at all levels of the organizational structure); complexity (all parameters of company activity are the objects of control); cross- and multi-level (the control over different parameters of the company activity is carried out from different positions, providing more qualitative control over threats to its financial stability). The aforementioned positively characterizes the effectiveness of the modern concept of internal control, however, significantly increases the cost of conducting a business, and weakens its competitiveness with FinTech companies. At the same time, it is not expected that quantitative and qualitative characteristics of regulatory norms will decrease in the near future.

As a result, financial institutions must find the ways to optimize the use of their limited resources using innovative technologies to build effective (both in terms of efficiency and the criterion of "revenues/effects/costs"), based on risk-oriented approaches, compliance systems. In our opinion, the only way out of this situation is a cooperation between traditional financial institutions with Infrastructure FinTech companies; first of all, we are talking about the use of RegTech companies' services and companies specializing in Institutional and B2B services. The Institute of International Finance defines RegTech as "the use of new technologies to solve regulatory and compliance burdens more effectively and efficiently".

That is, the use of innovative regulatory technology tools can help improve the quality and effectiveness of financial institutions' compliance systems that create preconditions for increasing the protection of depositors, lenders, investors. However, these tools may also raise new challenges and regulatory implications for financial institutions. It will be described further.

Considering the cooperation of traditional financial intermediaries and RegTech companies, we identify the following directions: diagnosis and monitoring; clients' identification and countering money laundering; regulatory norms and requirements;

reporting and risk management. Nowadays traditional financial intermediaries have real opportunities for optimization of compliance functions using such technologies as artificial intelligence (AI), natural language processing, big data and advanced analytics, cloud-based computing, robotics process automation, distributed ledger technology, application program interfaces (APIs) and biometrics [37].

The systematization of the current regulatory challenges of traditional financial intermediation and the tools of innovative regulatory technologies that can contribute to their solution in the most effective way is given in Table 2.

It made it possible to find out the following:

(1) Money mediation, the main objectives of which are to ensure proper profitability and qualitative satisfaction of the financial needs of economic agents, for the management of regulatory risks. Contemporary conceptual changes in the financial regulation system led to the transformation of banks into think tanks. Currently, a traditional financial intermediator is obliged to provide a deep, multi-dimensional, comprehensive study of the current functioning environment, and systematically simulate the potential level of risk practices in the

Table 2 Activity fields of traditional financial intermediaries requiring the use of technological tools

The most essential regulative changes in recent years	Challenges for financial institutions due to new/improved regulative norms	Tasks for innovative technological companies
Adoption of payment services directive (EU) 2015	Personal data fragility Risk of scammers Fraudulent activity	Consolidate multi-banked accounts information in 1 portal Create an algorithm that can guarantee high data security (data transfer, data usage)
Adoption of Basel III	Strengthening of risk management processing (on corporate level as well as on regulative one) Implementation of dynamics (e.g. for liquidity risk management) and countercyclical indicators	Data lineage Risk aggregation Workflow automation Validation and audit functionality and disclosure in multiple formats Complete data management, integration & warehousing platform for all financial data across the institution, including all subsidiaries Automatization of fund documentation management BigData analytics Biometric identification (e.g. facial recognition technology)
Implementation of IFRS 9	Need to use complicated complex models for estimating expected losses	
	Deeping of clients' identification	

Source Developed by the authors

event of a change in certain operating conditions, primarily external ones. It makes financial intermediaries continuously and systematically carry out the accumulation, systematization, and subsequent processing of a large array of disparate data. At the same time, the ability to do exactly this activity, in our opinion, becomes the main market barrier to enter this business sphere;

(2) The data on the environment in which the financial intermediators function, are no less important assets than cash. Moreover, the quality and efficiency of data management is becoming a fundamental ground of effective money mediation from the position of finite financial results;

(3) The use of RegTech firms' services by financial intermediaries is appropriate in the following areas: risk management; digital identity; fraud monitoring and control; data management; e-KYC (know your client); governance; compliance and regulatory reporting.

We consider the in-depth study of the possibilities of RegTech companies' services in terms of improving the efficiency of financial intermediary risk management systems to be the most reasonable (Table 3). The latter directly affects reducing their Basel III capital requirements. According to the European Central Bank in 2019 the three most prominent risk drivers affecting the euro area banking system are: geopolitical uncertainties, the stock of non-performing loans (NPLs) and potential build-up of future NPLs, and cybercrime and IT disruptions. These are followed by repricing in financial markets, low interest rate environment and banks' reaction to regulation [37].

Table 3 Features of the use of RegTech companies' services to improve financial intermediators' risk management systems

The type of risk	Scope of innovative technological companies' tools	Results
Operating risk	Real-time document management platform	Minimizing the negative effects of technical failures; leakage of information; reduction of internal information asymmetry (use of a single information base by all units of the system)
Credit risk	Credit scoring systems based on big data analytics technology;	The ability to consistently maintain the level of credit risk of retail lending
Liquidity risk	Machine learning Predictive analysis	It allows to provide multivariate diagnostics system of activity conditions and increase the efficiency of business decision making, response preventively to the most probable threats
Interest rate risk		

Source Developed by the authors

In general, today it is important to discuss the issues of the formation of RegTech ecosystem, which should provide such tasks (based on the approach developed by Deloitte [38]):

(1) compliance (34% significance of all Regtech functions) that includes: regulatory watch and online library of all regulative changes; project management; health check; web due diligence and security;
(2) identity management and control (24% of total significance);
(3) risk-management (20% of total significance) that means: scenario modelling and forecasting; risk exposure computation; risk reporting;
(4) reporting solutions (11% of total significance);
(5) transaction monitoring and audit systems (11% of total significance).

5 Financial Stability

In spite of the obvious benefits of innovative technological companies development and their increasing collaboration with traditional financial intermediation, the experience of the global crisis of 2008 proves that it is necessary to examine and identify potential channels of systemic threats to overall financial stability. Evolutionary transformations of financial mediation are evident nowadays (Table 4).

Of course, it will affect the concept of ensuring financial stability and the specificity of considering the peculiarities of financial innovations development. Modern tendency of financial innovations development is their high technological level. Therefore, it is important to determine the features of FinTech's impact on financial stability. We should admit that current state of FinTech does not reveal any significant signs of threats to financial stability.

However, if in future qualitative and quantitative parameters of their development (which is highly expected) are kept, FinTech can become a channel of systemic risks. That is why a special attention should be paid to the understanding of potential threats to financial stability because of FinTech active development. They require systematic monitoring by relevant providers of financial stability (Table 5).

The financial crisis of 2008 demonstrated the importance of crisis strategy for banks. It has become a new management ideology, which is aimed at substantiating and choosing different strategic approaches to the innovative development of banking institution, identifying the growing segments of the banking services market, optimizing costs and diversifying its activities. In particular, we see that traditional approaches of bank crisis management are mainly aimed at identifying and overcoming the crisis, while the innovative ones—on its prevention and ensuring post-crisis recovery and growth. Financial analytics systems (digital data management platforms); e-commerce and digital marketing systems; digital platforms for the formation of bank's resource base; digital technologies (AI, Big Data, mobile technologies, cloud computing, IOT, Blockchain, machine learning, etc.) are of interest in this aspect.

Table 4 Targets of financial innovations use by financial institutions in the context of diffusion channels

Diffusion channel	The purpose of using financial innovation	
	In the pre-crisis period	Under modern conditions
Distribution (S)	Maximizing of financial *products* sales volume	Maximizing the usefulness of financial *services* for clients based on FinTech
Risk management (Pr)	Maximizing profits, minimizing financial risks due to complex financial derivatives	Improving the efficiency of capital use
Financial instruments (P)	Development of financial instruments for eliminating initial risk by a financial agreement from a financial institution's balance sheet	Development of specialized financial instruments for risk management
Organization of internal business processes (O)	Reducing transaction costs, increasing the informational openness of financial intermediaries	Minimizing operational risks, reducing compliance costs, improving the transparency of financial intermediaries
Institutional infrastructure (O)	Active development of innovative forms of financial intermediation in securities market	Active development of FinTech in all business processes of financial intermediaries, strengthening the integrity of the financial system (P2P, B2B, P2B), increasing cqual acccss to financial services

P product; *Pr* process; *S* service; *O* organization
Source Developed by the authors

In recent years, AI has become popular in banking industry. Where it is used to automate customer service (virtual assistants or chat bots based on the technology of natural language processing), making sound and cost-effective decisions on lending, optimizing business processes—reducing operating costs and minimizing risks, introducing assistants for investing, security, image recognition and fraud prevention (Bank of America, UBS, JPMorgan Chase, OCBC Bank, Bank Shizuoka etc.).

6 Conclusion

Today, FinTech is a recognized driver of the evolutionary transformation of traditional financial intermediation, as it stimulates high pace of technological innovation, changing expectations and experience of financial services consumers. FinTech

Table 5 FinTech threats to financial stability

Content of threats to financial stability	The object of monitoring
Increasing of compliance costs of traditional financial intermediaries leads to active cooperation with FinTech companies on outsourcing terms for individual business areas and operational processes. Such process diversification results in delimitation of objects of regulatory influence within one financial service, and, therefore, reduces the limits of business responsibility	The dynamics of partnership between traditional financial institutions and FinTech companies
FinTech companies operate either outside the regulatory perimeter of traditional financial institutions or are subjects of substantially lower supervisory and regulatory standards Under the conditions of substantial growth of this business, the influence of regulatory arbitrage on the effectiveness of the supervision system that controls the activity of these firms can be considerably strengthened, that is, the control level of the financial system as a whole will deteriorate	The supervision and regulation system of the activities of non-bank credit institutions and FinTech companies, it compliance with the risk level of their activities, the depth of interconnectivity of FinTech companies and traditional financial institutions. Cross-border activity of FinTech
The deepening of network connections between different types of institutions in the field of financial services increases the sensitivity of the financial system to: (1) cyber-attacks; (2) risk of unreliability of the third party (the association of a particular set of financial institutions, especially in the case of systemically important, with one FinTech firm) may lead to operational failures; (3) reputation risks	The quality of cyber security systems; the level of FinTech market concentration in terms of specialization; the level of protection of FinTech companies' financial services consumers (FinTech lending, FinTech payments)
FinTech covering of a critical mass of economic agents demand for making payments and transfers. Operating malfunctions in payment services (due to objective reasons or deliberately organized) are expected to remove the economic system from equilibrium, worsen the business climate, reduce trust in the institution of financial intermediation in general	The activity of FinTech in the payment and money transfer market and the level of this segment concentration
Maximum implementation of AI to minimize the cost of services can stimulate the emergence of new and unpredictable sources of financial markets infection	Implementation of AI as an alternative to human resources

Source Developed by the authors on the basis of [39, 40]

offers a wide range of opportunities, including cost reduction, efficiency improve-ments, reduction of information asymmetry, and wide access to financial services, economic development and inclusive economic growth, simplify and improve the quality of compliance control and supervision.

Rapid changes at financial services market connected with FinTech has made tra-ditional financial intermediators seek a strategy for cooperation and build a partner-ship model, increase the level of adaptation to changes in the business environment, develop and implement innovative solutions to increase their own operational effi-ciency, develop new personalized financial products and services, create new chan-nels for interaction with customers. Therefore, not only interaction with the FinTech ecosystem is important, but also the transformation of financial intermediaries into ecosystem of business models is of great importance. At the same time, the suc-cess of FinTech development depends to a certain extent on financial intermediators' investments.

Recognizing the positive effects of FinTech, the latter are considered not only as a competitive threat, but also as a factor in increasing the risks of violating the rights of consumers and investors, the loss of effectiveness of the legal framework, the integrity and financial stability of the financial system in the process of adaptation. At the same time, it is important to adapt the regulatory system to ensure a balance between the introduction of financial innovation, economic growth and financial stability. It is reasonable to use the services of RegTech companies, because their technological solutions give a possibility to build a comprehensive system of financial intermediators' management, based on the principles of multi-aspect, continuality, interconnectedness and interdependence, security, and priority of security. On the one hand, the collaboration between RegTech and the financial intermediaries allows the latter to optimize resource costs when performing compliance functions. And on the other hand, the application of innovative regulatory technologies strengthens the internal integrity of institutions, increases their strength and resistance to shocks, greatly enhances the effectiveness of regulatory norms and requirements, improves the quality of risk-based supervision and monitoring flexibility, reduces opportunities for fraud and other types of criminal activity. It means that between two banks that are characterized by identical values of indicators of economic norms and requirements of regulatory bodies a bank that implements the services of RegTech companies is more financially stable and has a higher level of financial security.

In our opinion, the scenario for the development of financial intermediation, in which the use of various FinTech tools will become a kind of substitute for the traditional regulatory system, is quite possible. It will be the subject of our further research.

References

1. Kaczor S, Kryvinska N (2013) It is all about services—fundamentals, drivers, and business models. J Serv Sci Res 5(2):125–154 Springer, The Society of Service Science
2. Kryvinska N (2012) Building consistent formal specification for the service enterprise agility foundation. J Serv Sci Res 4(2):235–269 Springer, The Society of Service Science
3. Gregus M, Kryvinska N (2015) Service orientation of enterprises—aspects, dimensions, Technologies. Comenius University, Bratislava
4. Kryvinska N, Gregus M (2014) SOA and its business value in requirements, features, practices and methodologies. Comenius University, Bratislava
5. Molnar E, Molnar R, Kryvinska N, Gregus M (2014) Web intelligence in practice. J Serv Sci Res 6(1):149–172 Springer, The Society of Service Science
6. Kirichenko L, Radivilova T., Zinkevich I (2017) Forecasting weakly correlated time series in tasks of electronic commerce. In: Proceedings of 2017 12th international scientific and technical conference on computer sciences and information technologies (CSIT), Lviv, pp 309–312. https://doi.org/10.1109/stc-csit.2017.8098793
7. Kirichenko L, Radivilova T, Zinkevich I (2018) Comparative analysis of conversion series forecasting in e-commerce tasks. In: Shakhovska N, Stepashko V (eds) Advances in intelligent systems and computing II. CSIT 2017. Advances in intelligent systems and computing, vol 689. Springer, Cham, pp 230–242. https://doi.org/10.1007/978-3-319-70581-1_16
8. Dobrynin I, Radivilova T, Maltseva N, Ageyev D (2018) Use of approaches to the methodology of factor analysis of information risks for the quantitative assessment of information risks based on the formation of cause-and-effect links. In: Proceedings of the 2018 international scientific-practical conference problems of infocommunications. Science and technology (PIC S & T). Kharkiv, Ukraine, pp 229–232. https://doi.org/10.1109/infocommst.2018.8632022
9. Kirichenko L, Bulakh V, Radivilova T (2017) Fractal time series analysis of social network activities. In: Proceedings of the 2017 4th international scientific-practical conference problems of infocommunications Science and technology (PIC S & T). Kharkov, Ukraine, pp 456–459 (2017). https://doi.org/10.1109/infocommst.2017.8246438
10. Kirichenko L, Radivilova T, Tkachenko A (2019) Comparative analysis of noisy time series clustering. In: Proceedings of the 3rd international conference on computational linguistics and intelligent systems (COLINS-2019), vol I. Kharkiv, Ukraine, Apr 18–19, pp 184–196
11. Ageyev D (2010) NGN network planning according to criterion of provider's maximum profit. In: Proceeding of the IEEE international conference on modern problems of radio engineering, telecommunications and computer science (TCSET-2010), pp 256–256
12. Ageyev D, Al-Ansari A, Qasim N Multi-Period LTE RAN and Services Planning for Operator Profit Maximization. In: Proceeding of the IEEE the experience of designing and application of CAD systems in microelectronics, pp 25–27. https://doi.org/10.1109/cadsm.2015.7230786
13. Ageyev D, Wehbe F (2013) Parametric synthesis of enterprise infocommunication systems using a multi-layer graph model. In: Proceeding of the IEEE 23rd international crimean conference microwave & telecommunication technology, pp 507–508
14. Al-Dulaimi A, Al-Dulaimi M, Ageyev D (2016) Realization of resource blocks allocation in lte downlink in the form of nonlinear optimization. In: Proceeding of the IEEE 13th international conference on modern problems of radio engineering, telecommunications and computer science (TCSET), pp 646–648. https://doi.org/10.1109/tcset.2016.7452140
15. Schindler J (2017) FinTech and financial innovation: drivers and depth. finance and economics discussion series 081. https://doi.org/10.17016/feds.2017.081
16. Darolles S (2016) The rise of fintechs and their regulation. Financ Stab Rev Banque de France (20):85–92
17. Bunea S, Kogan B, Kund A-G, Stolin D (2018) Fintech and the banking bandwagon. J Financ Transform 47:25–33
18. Cortina Lorente J, Schmukler S (2018) The Fintech revolution: a threat to global banking? World Bank Res Policy Briefs 4 (2018)

19. Treleaven P (2015) Financial regulation of FinTech. J Financ Perspect FinTech 3(3)
20. Tsai K (2017) FinTech and financial inclusion in china. report, Chinese Academy of Financial Inclusion, Renmin University of China
21. Creehan S, Borst N (2017) Asia's Fintech revolution. Federal Reserve Bank of San Francisco, Asia Focus
22. Demertzis M, Merler S, Wolff G (2018) Capital markets union and the Fintech opportunity. J Financ Regul 4(1):157–165. https://doi.org/10.1093/jfr/fjx012
23. The Pulse of Fintech (2018) In: Biannual global analysis of investment in fintech. KPMG International Cooperative
24. Fintech Trends to Watch (2019) In: Research report, CB Insights
25. Global RegTech continues to grow, with over $500 m invested in Q1 2018. FinTech Global (2018)
26. Capital invested in global RegTech companies has increased more than nine-fold over the past six years. FinTech Global (2018)
27. Pantielieieva N, Krynytsia S, Khutorna M, Potapenko L (2018) FinTech, transformation of financial intermediation and financial stability. In: Proceedings of IEEE 5th international scientific-practical conference problems of infocommunications. Science and technology (PIC S & T 2018). Kharkov, Ukraine, pp 553–560. https://doi.org/10.1109/infocommst.2018.8632068
28. FinTech, RegTech and SupTech (2017) What they mean for financial supervision. Toronto Leadership Centre, Toronto
29. Global venture capital investment in Fintech industry set record in 2017. Accenture Analysis Finds, Accenture Newsroom (2018)
30. Basel III: a global regulatory framework for more resilient banks and banking systems. Basel Committee on Banking Supervision (2011)
31. Basel III: finalising post-crisis reforms. Basel Committee on Banking Supervision (2017)
32. Strategy of the National Bank of Ukraine. Action Program 2019, National Bank of Ukraine (2019)
33. Baxter G, Rengarajan S (2017) The March towards digital money: bringing the underbanked in from the cold. Imperial College, Citigroup, London
34. Ukrainian venture capital & private equity overview 2017. Ukrainian Venture Capital & Private Association (2018)
35. The future of regulatory productivity, powered by RegTech. In: Financial services. RegTech position paper, Deloitte (2016)
36. Skinner C (2019) Compliance will kill the bank, Chris Skinner's Blog
37. ECB banking supervision: risk assessment for 2019. European Central Bank (2018)
38. Hugé F-K (2018) Using Regtech to transform compliance and risk support functions into business differentiators. Inside magazine, Deloitte
39. Financial stability implications from FinTech: supervisory and regulatory issues that merit authorities. Financial Stability Board (2017)
40. Wyman O (2017) Balancing financial stability, innovation, and economic growth. White paper. World Economic Forum (2017)

Data-Centric Formation of Marketing Logistic Business Model of Vegetable Market Due to Zonal Specialization

Volodymyr Pysarenko ⓘ, Olena Ponochovna ⓘ, Maria Bahorka ⓘ
and Volodymyr Voronyansky ⓘ

Abstract Data processing tasks provided by agrarian enterprises providing logistic and marketing of research as an actual methodological material for the formation of new business model of vegetable market Ukraine. The variant of products transportation, finding optimal supplies of products from the manufacturer to the consumers is determined in the article. The necessity of using data processing systems in analyzing the part of the food market of Ukraine from vegetable production is analyzed in this research. The analysis of trends in production and consumption of vegetable products in the regions of Ukraine on the basis of open source data was made. The forecast of volumes of vegetable categories by regions up to 2030 and the satisfaction of the demand for vegetables in the region by regional producers was made. It was discovered that there is a shortage in the Kyiv and Ivano-Frankivsk regions. The supply of missing products from neighboring regions from the point of view of transportation costs is substantiated. To solve the problem of optimizing the location of the center, according to the proposed criterion, the center of gravity method was used. As a result, the geographic coordinates of the warehouses, in which the total annual transportation costs of the cargo in the considered system will be minimal, are calculated.

Keywords Management and marketing logistic · Data-Centric business · Model · Vegetable market

V. Pysarenko (✉) · O. Ponochovna
Poltava State Agrarian Academy, Poltava, Ukraine
e-mail: volodymyr.pysarenko@pdaa.edu.ua

O. Ponochovna
e-mail: elena.ponotchovna@gmail.com

M. Bahorka
Dnipropetrovsk State Agrarian and Economic University, Dnipro, Ukraine
e-mail: masha010574@gmail.com

V. Voronyansky
Poltava Oil and Gas College of National Technical Yuri Kondratyuk University, Poltava, Ukraine
e-mail: volodimir.voronjanski@gmail.com

© Springer Nature Switzerland AG 2020   23
D. Ageyev et al. (eds.), *Data-Centric Business and Applications*,
Lecture Notes on Data Engineering and Communications Technologies 42,
https://doi.org/10.1007/978-3-030-35649-1_2

1 Introduction

The objects of the food market infrastructure include wholesale and retail enterprises, auctions, fairs, exhibitions, commodity exchanges, communication systems, state institutional structures, etc.

Providing trade, marketing, logistics, information services, as well as services in storage, transportation and packaging of products, objects of the food market infrastructure should contribute to creating the necessary conditions for the effective operation of market entities and timely provision of food products to the population [1]. Data processing tasks provided by these enterprises provide logistic, marketing, economic and other fields of research as an actual methodological material for the formation of new strategies and correction of existing ones [2, 3].

Despite some positive changes, the level of infrastructure development remains a deterrent to the functioning of the market [4]. In the system of commodity sales of agricultural products, the organization of the functioning of wholesale markets needs to be improved; information provision of market participants is inadequate, and the data they submit requires further processing.

In order to improve the functioning of the food market of the region and to establish stable relations between commodity producers and consumers, the creation of a network of inter-regional wholesale food markets, which will promote the effective promotion of food products to the final consumer, should be considered as a priority [5]. Solving this problem requires the processing of data of different classes (geographical coordinates [6], distances, volumes of supplying goods [7], prices for goods [8], etc.) with the help of information systems [9, 10]. It is necessary to justify the choice of both the information system [11] and the methods of data processing in it [12].

Effective system of distribution of vegetable products through wholesale markets, firstly, guarantees to agricultural producers and processing enterprises equal conditions of sales of products; and secondly, it provides the supply of food for the population of cities and large settlements during the year; thirdly, weakening the dependence of agricultural products on the monopoly of processing enterprises; and finally provides support for the national commodity producer.

In addition, the organization and functioning of the broad-based wholesale market in Ukraine [13] will significantly weaken the activities of intermediary structures, accelerate the advancement of trade flows, reduce losses and preserve the quality of vegetable products.

The presence of numerous sales channels does not always provide a positive result, enterprises can focus on contractual relations, a stable market.

Determining the variant of products transportation, finding optimal deliveries of products from the manufacturer to the consumers processing enterprises, wholesalers and retailers will allow reducing costs and increase profitability [9, 14].

This work is structured as follows. The introduction analyzes the necessity of using data processing systems in analyzing the part of the food market of Ukraine from vegetable production. In the first section, on the basis of open source data, an analysis of trends in production and consumption of vegetable products in the regions of Ukraine was made. On the basis of the conducted analysis, the forecast of volumes of vegetable categories by regions up to 2030 and the satisfaction of the demand for vegetables in the region by regional producers was made. It was discovered that there is a shortage in the Kyiv and Ivano-Frankivsk regions. The second section deals with the planning of regional distribution centers for the supply of intra-regional consumption. The supply of missing products from neighboring regions from the point of view of transportation costs is substantiated. To solve the problem of optimizing the location of the center, according to the proposed criterion, the center of gravity method was used. As a result, the geographic coordinates of the warehouses, in which the total annual transportation costs of the cargo in the considered system will be minimal, are calculated. For this purpose, data types such as geographic coordinates of suppliers and their volumes of traffic for a year have been used. Section 4 concludes and describes future steps.

2 Data-Centric Forecast of Vegetable Production Within Regions for Business Information Processing and Management

The vegetable market should be distinguished as a separate, independent one. The reason for this is a large total volume of manufactured goods and a part in the production of all regions of Ukraine. After the transition of Ukrainian agriculture to market relations in trade of vegetable products, certain changes took place.

The cancellation of the state order and the suspension the operation of the procurement network of consumer cooperatives for the objective reasons led to a sharp decrease in the volumes of production and sale of vegetable products by large specialized enterprises, as well as significant structural changes in production and production sales channels, and there was a tendency to reduce purchases of vegetable products by procurement organizations and increased sales of vegetables on the market, to commercial structures [15].

Vegetable producers have some difficulties with the marketing of grown products. The implementation of these products is significantly complicated by the lack of small commodity formations, which on a contractual basis could guarantee the supply of a certain volume of products to wholesale markets or processing enterprises. Instead, most intermediaries raise sales prices or supply imported products to the market, artificially reducing the access of most people to vegetables consumption.

Since the development of vegetable growing significantly depends on the deepening of its agro-industrial integration with industrial processing of vegetables and the expansion of processing and storage of products in the places where they grow, they should study carefully the possibilities of rational use and increase of capacities of the canning industry, strengthening the material and technical base of subsidiary industries of farms and ensuring their vegetable stores, certain changes in the structure of production [16].

At this stage of development of vegetable products market, an important element of forecasting is also the definition of promising yields. As you know, the level of productivity of vegetable crops depends on many factors: soil and climatic conditions, technologies of cultivation and storage of crop, varietal composition, etc. It should be noted that these factors are interrelated and operate in the complex.

In the conditions of limit and sharp increase on prices for resources, one of the main factors in ensuring the maximum return on investment and the necessary rates of increase in volumes and achievement of the stability of the production of vegetable products is to improve the territorial organization of the industry in the system of the agro-industrial complex of Ukraine.

Data processing in the regions of the country showed a significant differentiation of indicators of production and consumption of vegetables in general in Ukraine [17]. Satisfaction of the offer on the market of vegetable products within separate regions is considerably different, and these differences increase with years.

At the same time, there is a situation in which the high level of vegetables consumption in some cases corresponds to a relatively low level of production [18].

The solution of this problem can be achieved by using the optimal distribution of vegetable production in the natural and climatic zones of Ukraine, taking into account the production volumes of the main types of vegetables and the necessary minimum amount of their consumption.

In our opinion, the target function may be the minimum cost of transportation of vegetable products. However, for its calculation it is necessary to forecast the supply volumes and consumption volumes of vegetable products for the current and subsequent years.

As it was already noted, the vegetable production gradually shifted from the Steppe and Forest-Steppe zones to Polissya and the Carpathians zones (Table 1).

As variables volumes of vegetable sales for the main consumption are defined:

X_{SS}—realization in the Steppe zone of the products produced in the zone of the Steppe;

X_{SF}—realization in the Steppe zone of products produced in the Forest-Steppe zone;

X_{SP}—realization in the Steppe zone of products produced in the Polissya zone;

X_{SC}—realization in the Steppe zone of products produced in the Carpathian zone;

X_{FS}—realization in the Forest-Steppe zone of products, produced in the Steppe zone;

X_{FF}—realization in the Forest-Steppe zone of products produced in the Forest-Steppe zone;

X_{FP}—realization in the Forest-Steppe zone of products, made in the Polissya zone;

X_{FC}—realization in the Forest-Steppe zone of products, made in the Carpathian zone;

X_{PS}—realization in the Polissya zone of products, produced in the Steppe zone;

X_{PF}—realization in the Polissya zone of products, produced in the Forest-Steppe zone;

X_{PP}—realization in the Polissya zone of products, produced in the Polissya zone;

Table 1 Summary table of zonal production of vegetable products in Ukraine (average for 2010–2018), Kt [19]

Index	Cabbage	Cucumbers	Tomatoes	Table beets	Carrots	Onions
Production, total	**1556.8**	**591.9**	**1137.9**	**580.6**	**505.9**	**556.9**
Including zones						
Steppe	558.1	90.8	809.3	168.4	135.6	268.9
Forest-Steppe	502.3	239.6	241.8	221.2	210.8	192.5
Polissya	348.8	222.4	46.5	149.8	132.7	72.4
Carpathians	147.6	39.1	40.3	41.2	26.8	23.7
Need, total	**1381.6**	**460.5**	**1796.1**	**506.6**	**460.5**	**460.5**
Including zones						
Steppe	601.1	200.4	781.4	220.4	200.4	200.4
Forest-Steppe	487.5	162.5	633.7	178.7	162.5	162.5
Polissya	214.3	71.4	278.6	78.6	71.4	71.4
Carpathians	78.7	26.2	102.4	28.9	26.2	26.2

X_{PC}—realization in the Polissya zone of products, produced in the Carpathian zone;
X_{CS}—realization in the Carpathian zone of products, produced in the Steppe zone;
X_{CF}—realization in the Carpathian zone of products, produced in the Forest-Steppe zone;
X_{CP}—realization in the Carpathian zone of products, produced in the Polissya zone;
X_{CC}—realization in the Carpathian zone of products, produced in the Carpathian zone.

As a result of the performed calculations (Table 2), the data on optimal volumes of sales of the main types of vegetable crops in the natural and climatic zones of Ukraine were obtained, taking into account the volumes of their production and the minimum required consumption volumes. Taking into account the data of the conducted research, the lack of production of tomatoes was found, which does not allow to sufficiently satisfy the needs of the population of the country (the deficit is 658.2 Kt). However, in the production of other types of vegetable products there is a surplus in relation to the needs of the population. Thus, it's a cabbage—175.2; cucumbers—131.4; table beets—73.9; carrots—45.4; onions—97.0 Kt. Therefore, we consider it expedient to satisfy the lack of tomatoes for the subsequent periods at the expense of imports, and for future periods to plan (predict) higher production volumes.

Predicted volumes of vegetables production and consumption in the nature and climatic zones of Ukraine, taking into account the dynamics of population and average consumption of vegetable products, are given in Table 3.

Using the optimal distribution of vegetable production in the natural and climatic zones of Ukraine, taking into account the production volumes of the main types of vegetables (cabbage, cucumbers, tomatoes, table beets, carrots and onions), and the

Table 2 The matrix of redistribution of vegetable production by natural and climatic zones of Ukraine, (on average for 2010–2019), Kt

Index	Cabbage	Cucumbers	Tomatoes	Table beets	Carrots	Onions
Steppe–Steppe	558.1	90.8	781.4	168.4	135.6	200.4
Steppe–Forest Steppe	–	–	27.9	–	–	–
Steppe–Polissya	–	–	–	–	–	–
Steppe–Carpathians	–	–	–	–	–	–
Forest-Steppe–Steppe	14.8	77.1		42.5	48.3	
Forest-Steppe–Forest-Steppe	487.5	162.5	214.8	178.7	162.5	162.5
Forest-Steppe–Polissya	–	–	–	–	–	–
Forest-Steppe–Carpathians	–	–	–	–	–	2.5
Polissya–Steppe	–	32.5	–	–	16.5	–
Polissya–Forest-Steppe	–		–	–		–
Polissya–Polissya	214.3	71.4	46.5	78.6	71.4	71.4
Polissya–Carpathians	–	–	–	–	–	–
Carpathians–Steppe	28.2	–	–	9.5	–	–
Carpathians–Forest-Steppe	–	–	–	–	–	–
Carpathians–Polissya	–	–	–	–	–	–
Carpathians–Carpathians	78.7	26.2	40.3	28.9	26.2	23.7
Surplus	175.2	131.4		73.9	45.4	97.0
Lack	–	–	658.2	–	–	–
Cost of redistributed production, ths. UAH	84288	688290	179955	119534	215748	5750
Economic effect, ths. UAH	46999	283584	527	58625	119856	1815

minimum amount of their consumption, the sizes of optimum sales volumes of the main types of vegetable crops are prognosed according to natural and climatic zones of Ukraine for 2020–2030, taking into account the volumes of their production and the minimum-necessary consumption volumes (Table 4).

Taking into account the research data, the sizes of optimum volumes of realization of the main kinds of vegetable crops according to the natural and climatic zones are prognosed. As a result of the calculations, there was a shortage in the production of tomatoes, which in 2020 will be 707.1 Kt, and in 2030—699 Kt, which does not allow to adequately satisfy the needs of the country population. However, in the production of other types of products, their surplus is observed in 2020–2030 relative to the needs of the population. Therefore, we consider it expedient to satisfy the lack of tomatoes for the subsequent periods at the expense of imports and to plan larger production volumes. The surplus of vegetable products can be rationally exported to countries where their shortage is observed and, therefore, the realization at favorable prices, for example, to the northern regions of the EU or The Baltic countries is possible.

Table 3 The size of the territory, population density and vegetable production zones (segments) in Ukraine (average forecast for 2015–2030)

Nature and climatic zone	Territory, ths. km²	Population, ths persons (on the average)		Production fund total, Kt		Per capita, kg		Consumption fund total, Kt	
		2020	2030	2020	2030	20202	2030	2020	2030
Cabbage									
Steppe	250.2	19732.8	19454	568.3	589.5	28.8	30.3	647.2	669.2
Forest-Steppe	202.9	15421.8	15204	505.8	524.5	32.8	34.5	505.8	523
Polissya	123.8	6844.2	6747.5	348.4	361	50.9	53.5	224.5	232.1
Carpathians	26.7	2444.3	2409.8	145.7	151.1	59.6	62.7	80.2	82.9
Total	603.8	44443.4	43815	1568.9	1629.9	35.3	37.2	1457.7	1507.2
Cucumbers									
Steppe	250.2	19732.8	19454	248.6	256.8	12.6	13.2	217.1	223.7
Forest-Steppe	202.9	15421.8	15204	237.5	246.3	15.4	16.2	169.6	174.8
Polissya	123.8	6844.2	6747.5	78.7	79.6	11.5	11.8	75.3	77.6
Carpathians	26.7	2444.3	2409.8	32	33.3	13.1	13.8	26.9	27.7
Total	603.8	44443.4	43815	582.2	600.3	13.1	13.7	488.9	503.9
Tomatoes									
Steppe	250.2	19732.8	19454	818.9	846.2	41.5	43.5	818.9	832.6
Forest-Steppe	202.9	15421.8	15204	239	247.8	15.5	16.3	640	650.7
Polissya	123.8	6844.2	6747.5	41.7	43.2	6.1	6.4	284	288.8
Carpathians	26.7	2444.3	2409.8	37.6	39	15.4	16.2	101.4	103.1
Total	603.8	44443.4	43815	1137.8	1178.6	25.6	26.9	1844.4	1875.3
Onions									
Steppe	250.2	19732.8	19454	270.3	280.1	13.7	14.4	217.1	223.7

(continued)

Table 3 (continued)

Nature and climatic zone	Territory, ths. km²	Population, ths persons (on the average)		Production fund total, Kt		Per capita, kg		Consumption fund total, Kt	
		2020	2030	2020	2030	2020	2030	2020	2030
Forest-Steppe	202.9	15421.8	15204	195.9	202.2	12.7	13.3	169.6	174.8
Polissya	123.8	6844.2	6747.5	69.1	71.5	10.1	10.6	75.3	77.6
Carpathians	26.7	2444.3	2409.8	18.8	19.5	7.7	8.1	26.9	27.7
Total	603.8	44443.4	43815	546.7	574	12.3	13.1	488.9	503.9
Table beets									
Steppe	250.2	19732.8	19454	163.8	169.2	8.3	8.7	236.8	243.2
Forest-Steppe	202.9	15421.8	15204	220.5	229.6	14.3	15.1	185.1	190
Polissya	123.8	6844.2	6747.5	147.2	152.5	21.5	22.6	82.1	84.3
Carpathians	26.7	2444.3	2409.8	37.2	38.8	15.2	16.1	29.3	30.1
Total	603.8	44443.4	43815	586.7	604.6	13.2	13.8	533.3	547.7
Carrots									
Steppe	250.2	19732.8	19454	130.2	138.1	6.6	7.1	213.1	221.8
Forest-Steppe	202.9	15421.8	15204	211.3	215.9	13.7	14.2	166.6	173.3
Polissya	123.8	6844.2	6747.5	130.7	135.6	19.1	20.1	73.9	76.9
Carpathians	26.7	2444.3	2409.8	22.2	22.9	9.1	9.5	26.4	27.5
Total	603.8	44443.4	43815	493.3	512.6	11.1	11.7	480	499.5

Table 4 The matrix of redistribution of vegetable products according to the natural and climatic zones of Ukraine (forecast for 2020–2030), Kt

Index	Cabbage	Cucumbers	Tomatoes	Table beets	Carrots	Onions
2020						
Production, total	1568.2	596.8	1137.2	568.7	494.4	554.1
Need, total	1457.7	488.9	1844.3	533.3	480	488.9
Realization according to zones						
Steppe–Steppe	568.3	75.3	41.7	163.8	130.2	18.8
Steppe–Forest-Steppe	–	–	–	–	–	–
Forest-Steppe–Steppe	–	–	–	35.4	22.2	–
Forest-Steppe–Forest-Steppe	505.8	217.1	37.6	185.1		69.1
Forest-Steppe–Polissya	–	–	–	–	–	–
Forest-Steppe–Carpathians	–	–	–	–	–	–
Polissya–Steppe	78.9	–	–	–	44.7	–
Polissya–Forest-Steppe	–	–	–	–	–	–
Polissya–Polissya	224.5	26.9	239	29.3	166.6	169.6
Carpathians–Steppe	–	–	–	–	16	
Carpathians–Forest-Steppe	–	–	–	–	26.4	6.2
Carpathians–Polissya	–	–	–	–	–	–
Carpathians–Carpathians	80.2	169.6	818.9	82.1	73.9	217.1
Surplus	110.5	107.9		35.4	14.4	65.2
Lack	–	–	707.1	–	–	–
Cost of sold products, ths. UAH	1497.15	488.9	1137.2	560.95	522.5	494.5
2030						
Production, total	1626.1	616	1176.2	590.1	512.5	573.3
Need, total	1507.2	503.8	1875.2	547.6	499.5	503.8
Realization according to zones						
Steppe–Steppe	589.5	77.6	43.2	169.2	138.1	19.5
Steppe–Forest-Steppe	–	–	–	–	–	–
Steppe–Polissya	–	–	–	–	–	–
Steppe–Carpathians	–	–	–	–	–	–
Forest-Steppe–Steppe	1.5	–	–	74	22.9	8.2
Forest-Steppe–Forest-Steppe	523	223.7	39	155.6		63.3
Forest-Steppe–Polissya	–	–	–	–	–	–
Forest-Steppe–Carpathians	–	–	–	–	–	–
Polissya–Steppe	–	–	–	–	42.6	–
Polissya–Forest-Steppe	–	–	–	8.7		14.3

(continued)

Table 4 (continued)

Index	Cabbage	Cucumbers	Tomatoes	Table beets	Carrots	Onions
Polissya–Polissya	232.1	27.7	247.8	30.1	173.3	174.8
Polissya–Carpathians	–	–	–	–	–	–
Carpathians–Steppe	–	–	–	–	18.2	–
Carpathians–Forest-Steppe	–	–	–	25.7	27.5	–
Carpathians–Polissya	–	–	13.6	–	–	–
Carpathians–Carpathians	82.9	174.8	832.6	84.3	76.9	223.7
Surplus	118.9	112.2	–	42.5	13	69.5
Lack	–	–	699	–	–	–
Cost of sold products, ths. UAH	1546.67	503.8	1179.6	574.7	542.5	509.43

At the same time, the share of large enterprises in the total production volume is decreasing. There was also a decrease in the crop area by 40 ths, ha, gross yield—by 2 Mt, yields—by 22%. Reduced yields are decreasing both for large and small farms [20].

The supply of vegetables in Ukraine is formed, mainly due to production, which is concentrated in households (Fig. 1).

In 2018, there continues to be a positive tendency of increase in production volumes, which is likely to continue in the near future (Fig. 2). The reason for this is the dynamic development of the vegetable market in Ukraine in recent years, which is still not saturated. The high profitability of this industry for the correct approach to growing and marketing, even in the overproduction season, in contrast to other segments of agro-industrial complex can be indicated as a distinctive feature.

In reference to the forecast of the vegetable subcomplex development, structural changes in production are expected: the share of specialized agricultural enterprises

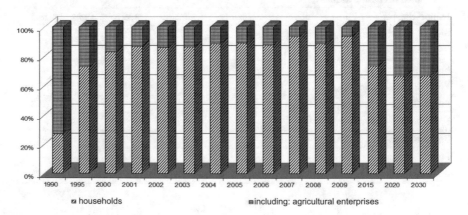

Fig. 1 The share of agricultural enterprises and households in the vegetables production, 1990–2030

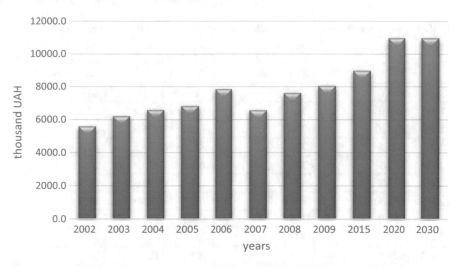

Fig. 2 The dynamics of vegetable production, 1990–2030

and peasant farming will increase, and households will decrease. There was a tendency towards a decrease in the purchase of vegetable products by stock organizations and increased sales of vegetables on the market, to commercial structures, as well as on payroll. The prices for vegetable products did not contribute to the expanded reproduction of production and did not compensate for production costs.

The economic crisis has led to a violation of the interaction of vegetable subcomplex individual spheres: production, processing, storage and marketing of vegetable products, the processes of self-regulation and self-survival of its spheres began. In these conditions, the share of households in the vegetables' production has increased and their production in agricultural enterprises has decreased. In recent years, there has been a tendency to increase the production of vegetables in households (up to 90%) and decrease in agricultural enterprises (Table 5). At the same time, vegetable farming in farms has not yet become a high-value industry. One of the directions in its development may be the creation of specialized farms with fluctuations in the area of vegetables from 20 to 100 ha. A specialized farm must be integrated with similar or processing ones.

Depending on the area of arable land on the household and the availability of sales market, the level of energy efficiency and direction of specialization, the total area under the vegetables and the structure of their planting are determined. For farms, different options for the development of vegetable production are acceptable, taking into account the specific conditions of production and the availability of market niches. As the foreign experience of farms activity shows, growing of 3–4 main vegetable crops or specializing in monoculture growing is common and effective.

In addition to vegetable production, in the Polissya and Forest-Steppe zones, formation of vegetable and potato farms, in the regions of the Steppe and Forest-Steppe—of specialized farms with the cultivation of a variety of raw materials for the

Table 5 Production of vegetables and melons in Ukraine by categories of farms, Kt, 1990–2030

Year	All categories of farms	Including: agricultural enterprises	Among them		The share of households in all categories of farms
			Farms	Households	
1990	6666.0	4872.0	0.0	1794.0	26.9
1995	5880.0	1607.0	27.4	4272.7	72.7
2000	5821.3	986.3	82.6	4835.0	83.1
2001	5906.8	772.5	90.7	5134.3	86.9
2002	5827.1	706.7	109.0	5120.4	87.9
2003	6538.2	827.6	131.7	5710.6	87.3
2004	6963.9	768.6	142.5	6195.3	89.0
2005	7295.0	780.7	156.1	6514.3	89.3
2006	8058.0	974.2	223.9	7083.8	87.9
2007	6835.2	713.4	168.2	6121.8	89.6
2008	7965.1	1108.6	275.9	6856.5	86.1
2009	8341.0	1120.1	223.7	7221.0	86.6
2015	10,000	1436	337	8564	85.6
2020	12,000	2069	387	9931	82.8
2030	13,000	2430	406	10,570	81.3

range of vegetable crops is possible. On average, over 20 years (2000–2018) in the structure of vegetable production 22% are different kinds of cabbage, 20% tomatoes, 20% beets and carrots, 12% onions (Fig. 3).

The largest production volumes of vegetables in the period 2000–2018 were reorded in Dnipropetrovska (506.9 Kt) and Khersonska (488.3 Kt) oblast (Fig. 4).

In the demand pattern for vegetable products, the main share falls on food— 72.1% and feed—14.3%. In the long term, by 2020, the share of volumes used for feed (up to 18%) is likely to increase, and losses will significantly decrease (from 6% in 2008 to 2% in 2020) (Fig. 5). At the same time, the physiological need for vegetable consumption will gradually be reached.

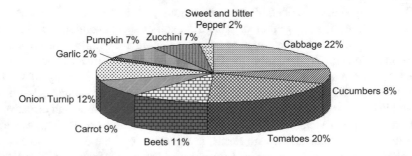

Fig. 3 Structure of production volumes of vegetables, 2000–2018

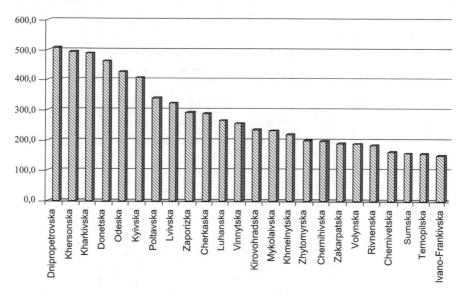

Fig. 4 Production volumes of vegetables in 2000–2018 in the regions, Kt

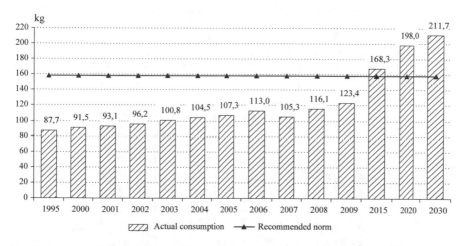

Fig. 5 Provision of population needs in vegetables per one person, kg/year, 1990–2030

In recent years there has been a tendency towards a decrease in the consumption of vegetables. So, in 1990, the production of vegetables per capita amounted to 141 kg, in 2000—91.9 kg, in 2010—only 101.7 kg. Reducing consumption is due to lower productivity, transformation of ownership in the state. Consumption of vegetables since 2000 increases, but the recommended rate—158 kg per person per year—should be achieved in 2020.

In recent years, there has been a positive trend towards increase in the vegetable production in Ukraine (Table 6).

Table 6 Vegetable production in Ukraine in 2000–2018

Index	Year								
	2000	2001	2002	2004	2013	2015	2016	2017	2018
Gross yield, Kt									
Vegetables, total	6666.4	5879.8	5821.3	6963.9	7295	8058	6835.2	7965.1	8341
Open-ground vegetables	6416.1	5607.4	5584.5	6697.7	7018.5	7755.3	6547.5	7669.8	7967
Closed-ground vegetables	250.3	272.4	236.8	266.2	276.5	302.7	287.7	295.3	374
Yield, c/ha									
Vegetables, total	149	120.2	112.3	148.7	157.1	171.4	152.3	173.9	182.8
Open-ground vegetables	144.2	115.4	108.2	143.8	152	165.9	146.8	168.5	175.7
Closed-ground vegetables	4.8	4.8	4.1	4.9	5.1	5.5	5.5	5.4	7.1
Gathered area, ths.m²									
Vegetables, total	447.2	489.3	518.6	468.2	464.4	470.3	448.8	457.9	456.36
Open-ground vegetables	445.0	485.7	516	465.7	461.8	467.6	446.1	455.3	453.46
Closed-ground vegetables	2.2	3.6	2.6	2.5	2.6	2.7	2.7	2.6	2.9

The production of vegetables in all categories of Ukrainian farms has stabilized in recent years and is about 8 million tons per year. In 2018, the production of open-ground vegetables increased by 36.8% compared with 2000, the production of green peas increased 3 times, tomatoes—by 33.0%, cabbage—by 55.1%, beets—by 31.9%, onions—by 86.3%, carrots—by 49.2%.

In 2018, the production of almost all vegetable crops increased both in agricultural enterprises and in households [21].

Fluctuations in production and consumption of vegetables are correlated with changes in the balance of vegetables (Table 7).

In recent years, an increase in the consumption of vegetables has been observed, primarily due to the increase in their production, mainly in the private sector, and not at the expense of production in large agricultural enterprises, as the main source of income of vegetable products to the trading network. Consumption of vegetables per capita has grown. To a certain extent, this is the result of a reduction in the population. A higher level of production is responsible for a higher level of consumption of vegetable products per capita. Of course, there is a surplus of consumption over production.

As a result of processing data from [19], the deficit of production and supply of vegetable products by regions of Ukraine was projected. Calculations are presented in Table 8.

Table 7 Balance of vegetables (including canned and dried products in fresh weight), Kt

Index	Year							
	2000	2002	2005	2007	2017	2018	2020	2030
Production	5575.5	5607.9	6845.4	6585.3	7640.1	8341.0	12000.0	13000.0
Change in stock at the end of the year	180.9	−154.8	176.4	−76.5	620.1	534.0	590.0	630.0
Import	26.1	70.2	90.0	142.2	320.4	209.0	118.0	126.0
Export	27.0	31.5	135.0	268.2	225.9	312.3	814.0	1102.0
Spent on feed	655.2	912.6	1092.6	1025.1	1094.4	1200.0	1400.0	1600.0
Spent on planting	77.4	78.3	81.0	89.1	91.8	108.0	95.0	100.0
Losses	159.3	123.3	353.7	463.5	550.8	600.0	420.0	450.0
Consumption Fund	4501.8	4687.2	5096.7	4958.1	5377.5	5796.0	8799.0	9244.0
Calculated per person, kg	91.5	96.2	107.3	105.3	116.1	125.9	198.0	211.0
Number of population	49.2	48.7	47.5	47.1	46.3	46.0	44.4	43.8

Table 8 Determination of volumes of vegetables supplies by regions of Ukraine, 2018

Consumers	Deficit, Kt	Suppliers	Surplus, Kt	Volume share	Supply volumes, Kt	Economic effect, ths. UAH
Kyiv region	−130.5	Zhytomyr region	52.4	0.12	15.6	3029.3
		Vinnitsa region	66.8	0.15	19.6	22285.2
		Cherkasy region	130.6	0.29	37.8	34665.2
		Poltava region	175.3	0.39	50.9	20744.3
		Chernihiv region	23.4	0.05	6.5	3292.1
Total		–	448.5	1	130.5	84016.1
Ivano-Frankivsk region	−9.5	Zakarpatskyi region	92.2	0.28	2.6	1596.5
		Lviv region	100.9	0.31	3.0	1610.4
		Ternopil region	55.4	0.16	1.5	779.0
		Chernivtsi region	83.5	0.25	2.4	748.0
Total		–	332.0	1	9.5	4733.8
All	−140	–	780.5		140	88749.9

The deficit of vegetables production by regional farms og Kyiv and Ivano-Frankivsk regions, 130.5 Kt and 9.5 Kt respectively was revealed. This necessitates the redistribution of products at the expense of interregional supply of vegetables. At the same time it is necessary to consider additionally the solution of the problem of optimization of transport flows.

3 Justification of Interregional Supply of Vegetables on the Basis of Marketing Logistics with the Help of Data-Centric Applications

The minimum transport costs required to transport products from suppliers is offered as the main criterion for choosing the location of the warehouse. In [22–24], a solution of a similar class of network design problems was considered. In scientific works [25, 26] several methods of finding the location of the distribution warehouse are considered:

– full-fledged method. The task of choosing an optimal placement is solved by a complete overview and evaluation of all possible variants of distribution centers placement and performed on a computer by methods of mathematical programming;
– a heuristic method, based on human experience and intuition, in contrast to the formal procedure that forms the basis of a complete reboot method;
– method of determining the center of gravity. It is used to determine the location of one distribution center. The advantages of this method are the versatility and ease of use, the availability of data necessary for the calculation, accounting for the turnover of each supplier, the mathematical reasoning for determining the geographical coordinates of the distribution center, according to which the distance from suppliers to the composition will be proportional to the volume of freight traffic, which will reduce the total transport costs [27].

Coordinates of the gravity flows center, that is, the points around which it will be possible to place the distribution warehouse, are calculated according to the formulas:

$$x_0 = \frac{\sum_1^i q_i \times x_i}{\sum_1^i q_i}, \ y_0 = \frac{\sum_1^i q_i \times y_i}{\sum_1^i q_i} \tag{1}$$

where q_i—volume of traffic flow of i supplier, Kt/year;
x_i, y_i—coordinates of i supplier;
x_0, y_0—optimal coordinates of warehouse.

Since suppliers of vegetables—economic entities are not defined, it is proposed to take the extreme points of the boundaries of consumer regions with the boundaries of the supplier regions for the rationalization of intra-regional displacements of goods in terms of geographic coordinates.

The volume of goods shipped by each supplier will be directly proportional to the size of the surplus of the suppliers regions' production. Calculations are presented in Table 8.

We calculate the optimal value of the coordinates of the warehouse location—x_0, y_0—for the Kyiv region in accordance with (1). The results of calculations are formed in the Table 9.

That is, for such coordinates of the warehouses' location the total intra-regional transport costs will be minimal, although the terms of delivery are not known, but they are carried out from all neighboring regions.

It is proposed to use the Clark-Wright method [14] to construct an optimal supply scheme for vegetables from farms to distribution warehouse. The Clark-Wright

Table 9 Coordinates of warehouse placement in Ukrainian regions with deficit of vegetable products, 2018

Indexes	Geographic coordinates	
Region	Kyiv	Ivano-Frankivsk
X_0	31°24′	24°47′
Y_0	50°08′	48°59′

method is one of the approximate, iterative methods, and is intended to solve the problem of transportation by computer. The error of the solution does not exceed an average of 5–10%. The advantages of the method are its simplicity, reliability and flexibility, which allows taking into account additional factors that affect the final solution of the problem.

Initial data will be:

q_i—volumes of one supply of each farm, kg;

X_i, Y_i—suppliers coordinates (expressed in decimal fractions).

We will consider the schemes of vegetables transportation to the distribution center in two regions with a significant deficit of vegetable production by regional farms. Such regions are Kyiv and Ivano-Frankivsk with a deficit of 130.5 Kt and 9.5 Kt, respectively.

Supply of vegetables to the distribution center in the Kyiv region will be carried out by seven suppliers in accordance with the concluded contracts. In relation to the distribution center, they have the following location (Fig. 6).

Coordinates of the warehouse in the Kyiv region—$x_0 = 31.4$; $y_0 = 50.13$. The coordinates of the nearest point with the developed infrastructure are Pereyaslav-Khmelnitsky. Coordinates and volumes of suppliers' products are given in Table 10.

According to the original scheme, a separate route is organized for the delivery of the cargo to each individual recipient. Thus, the original traffic scheme includes only radial routes of the vehicle movement, and the number of radial routes coincides with the number of suppliers. In this case, the scheme of transportation consists of 7 radial routes.

The essence of the Clark-Wright method is that, proceeding from the original scheme of transportation, move step by step, to the optimal scheme of transportation with circular routes. For this purpose, such concept as a kilometer gain is introduced.

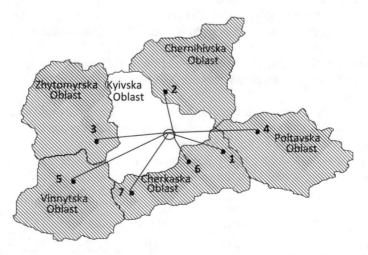

Fig. 6 Primary scheme of supply to the distribution center in the Kyiv region, 2018

Table 10 Coordinates and volumes of vegetable supply to the distribution center in the Kyiv region, 2018

#	Enterprise	Location	Geographic coordinates		Supply volumes q_i, Kt
			X_i	Y_i	
1	Vesna farms	Cherkasy region, Chernobayevsky district, Melnyky village, Lenina str., 90	31.25	49.28	11.5
2	Chereshenky Inc.	Chernihiv region, Koropsky district, Korop city, Dachna str., 25	30.9	51.02	34.5
3	Impak PI	Zhytomyr region, Andrushivsky district, Andrushevka city, Lysenko str., 25	28.82	50.1	17.3
4	Potatoes of Poltava region PI	Poltava region, Zinkovsky district, Birky village, Kuybysheva str., 52	32.75	50.23	28.8
5	Podillya PI	Vinnytsya region, Tyvrivsky district, Stroinytsi village, 50th anniversary of October str.	28.9	48.84	15.4
6	Demetra farms	Cherkasy region, Cherkasky region, Sagunovka village, Kosova str., 10	31.03	49.1	13.4
7	Obriy farms	Cherkasy region, Talnovsky region, Maydanetske village, Proletarian str., 31	30.2	48.3	9.6
	Total				130.5

For the scheme of transportation on radial routes, for example, 0-1-0 and 0-2-0, the total mileage of the vehicle is:

$$L_A = d_{01} + d_{10} + d_{02} + d_{20} = 2d_{01} + 2d_{02},$$

The scheme of transportation, which provides for the delivery of goods on the circular route, will be 0-1-2-0. Then the mileage of the vehicle will be:

$$L_B = d01 + d12 + d02.$$

The last scheme on the index of mileage of the vehicle, as a rule, provides a better result than the previous one. And so in the transition from radial routes to circular ones we get the next kilometer scheme:

$$s12 = LA{-}LB = d01 + d02{-}d12.$$

In total, the kilometer gain will be:

$$s_{ij} = d_{0i} + d_{0j}{-}d_{ij}, \tag{2}$$

where S_{ij}—kilometer gain obtained by combining points i and j, km;
d_{0i}, d_{0j}—distance between the wholesale base and the points i and j respectively, km;
d_{ij}—distance between points i and j, km.

We will calculate the distance between the points using the formula:

$$d_{ij} = \sqrt{(x_i - x_j)^2 + (y_i - y_j)^2} \tag{3}$$

For route 0-1-2-0:

$$d_{01} = \sqrt{(x_0 - x_1)^2 + (y_0 - y_1)^2} = \sqrt{(36.8 - 35.9)^2 + (48.53 - 48.6)^2} = 0.9;$$

$$d_{02} = \sqrt{(x_0 - x_2)^2 + (y_0 - y_2)^2} = \sqrt{(36.8 - 38.4)^2 + (48.53 - 48.23)^2} = 1.63;$$

$$d_{12} = \sqrt{(x_1 - x_2)^2 + (y_1 - y_2)^2} = \sqrt{(35.9 - 38.4)^2 + (48.6 - 48.23)^2} = 2.43.$$

Kilometer preference:

$$S_{12} = d_{01} + d_{02}{-}d_{12} = 0.9 + 1.63{-}2.43 = 0.1 \text{ (km)}.$$

In this case, as a result of organizing the scheme of circular delivery, the kilometer gain is insignificant and will amount to 0,1 km. Similar calculations will be made on other routes, where more significant gains are obtained.

The demonstration of using a step-by-step algorithm by the Clark-Wright method in relation to the considered problem is given in Table 11.

Step 1. On the matrix of kilometer preferences choose the cell (i*, j*) with a maximum kilometer preference S_{max}:

$$S_{max} = \max_{i,j} s(i, j) = s(i^*, j^*)$$

Table 11 The matrix of distances and kilometer preferences of transportation to the distribution center in the Kyiv region, 2018

	Matrix of distances between points (d_{ij}), km						
Matrix of kilometer preferences	0	0.9	1.63	1.24	1.92	1.62	0.94	2.14
(s_{ij}), km	0	1	2.43	0.35	2.68	2.52	0.3	2.88
	0	0.1	2	2.85	1.03	0.35	2.55	1.23
	0	1.79	0.2	3	2.94	2.86	0.54	3.13
	0	0.14	2.52	0.22	4	1.38	2.82	0.24
	0	0	2.9	0	2.16	5	2.5	1.58
	0	1.54	0.02	1.64	0.04	0.06	6	3.03
	0	0.16	2.54	0.25	3.82	2.18	0.05	7
	0	1.46	0.34	2.02	1	0.19	1.22	1.12
	0	0.08	2.93	0.14	3.72	2.6	0.01	4
	0	1.4	0	1.39	0.07	0.02	1.4	0.08

In doing so, the following three conditions must be met:

(1) points i* and j* are not part of the same route;
(2) points i* and j* are the starting and/or end point of those routes to which they are included;
(3) cell (i*, j*) is not blocked (that is, it was considered in the previous steps of the algorithm).

If we were able to install a cell that meets the three specified conditions, then proceed to step 2. If it fails, then proceed to step 6.

Step 2. The route, which includes the item i*, denote as route 1. Accordingly, the route, which includes the point j*, denote as a route 2. Introduce the following symbols: $N = \{1, 2, ..., n\}$—many suppliers; N_1 ($N_1 \subset N$)—subset of items that are part of the route 1; N_2 ($N_2 \subset N$)—subset of items that are part of the route 2.

Obviously that $i^* \in N_1$, $j^* \in N_2$ i $N_1 \cap N_2 = \emptyset$ (according to step 1, condition 1). Calculate the total volume of deliveries on routes 1 and 2:

$$q_1 = \sum_{k \in N_1} q_k \text{ i } q_2 = \sum_{k \in N_2} q_k$$

where q_k—supply volume of k point, kg (Table 5).

Step 3. Check the following condition:

$$q_1 + q_2 \leq c,$$

where c—cargo capacity of the vehicle, kg.

If the condition is satisfied, then proceed to step 4, if not—to step 5.

Step 4. Combine routes 1 and 2 into one general circular route X. We will assume that point i* is the end point of route 1, and point j* the starting point of route 2. When combining routes 1 and 2, we observe the following conditions:

– the sequence of points location on route 1 from the beginning and to the point i* does not change;
– point i* is associated with point j*;
– the sequence of points location on route 2 from point j* and to the end does not change.

Step 5. Repeat steps 1 through 4 until in the next repeat you can not find S_{max}, which satisfies three conditions in step 1.

Step 6. Count the total mileage of the vehicle.

Now we will consider the application of this algorithm to this problem.

The graphic representation of the problem is presented in Fig. 7.

According to the calculations, it is best to have two circular route of vegetables supply from suppliers to the distribution center: 0-3-5-7-0 and 0-6-1-4-0, and transportation from supplier 2 will be carried out along the radial route 0-2-0.

Ivano-Frankivsk region ranks third in the absence of demand for vegetables by domestic producers. Annual demand for vegetables from neighboring regions is 9.5 Kt.

Agreements on vegetables supply to the distribution center in the Ivano-Frankivsk region were concluded from eight suppliers every 3 days. Supplies are carried out by a truck of 3 tons, which belongs to the distribution center. We will plan the optimal way of vegetables transportation, if known suppliers'coordinates and volumes of one supply of each.

Coordinates of the distribution center in Ivano-Frankivsk region are $x_0 = 24.52$; $y_0 = 48.98$, volumes of suppliers' products are given in the Table 12. According to

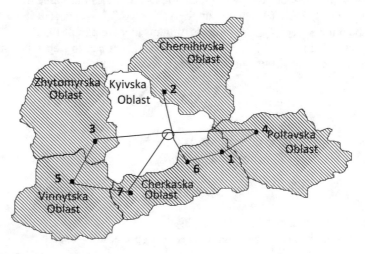

Fig. 7 Optimal scheme of transportation of vegetables to the distribution center in the Kyiv region, 2018

Table 12 Coordinates and volumes of vegetables supply to the distribution center in Ivano-Frankivsk region, 2018

#	Enterprise	Location	Geographic coordinates		Supply volumes q i, Kt
			X_i	Y_i	
1	Dream agroholding	Ternopil region, Gusyatinsky district, Vasylkivtsi village	25.21	49.05	1.9
2	Roveks Inc.	Ternopil region, Ternopil city, Podolska atr., 38	25.7	49.61	1.5
3	Blessing farms	Chernivtsi region, Novoselitsky district, Ryngach village	25.78	48.15	0.5
4	Zalissya farms	Lviv region, Brodovsky district, Brody city, Tudora str., 74	23.5	49.62	0.6
5	Oros farms	Zakarpatskyi region, Khustsky district, Iza village, Vakarova str., 10	22.35	48.7	1.5
6	Saldobosh farms	Zakarpatskyi region, Khustsky district, Steblivka village, Travneva str., 19	25.9	48.27	1.4
7	Source farms	Chernivtsi region, Vyzhnytsky district, Baniliv village	25.85	48.47	1.2
8	Agro+ farms	Lviv region, K.-Bugsky district, Remeniv village, Sadova str., 12	24.55	49.9	0.9
	Total				9.5

the given coordinates, it is expedient to place the distribution center in the city of Ivano-Frankivsk region.

Regarding the distribution center, suppliers have the position shown in Fig. 8.

Using the Clark-Wright method, by the formulas (4) and (5) we calculate the distance between suppliers and kilometer gain, obtained by combining points during transportation of vegetables to the distribution center.

Thus, according to the calculations, the best three circular routes for the transportation of vegetables from suppliers to the distribution center in Ivano-Frankivsk region: 0-4-5-6-0, 0-3-7-2-0 and 0-1-8-0.

The optimal scheme of transportation of vegetables to the distribution center in the Ivano-Frankivsk region is shown in Fig. 9.

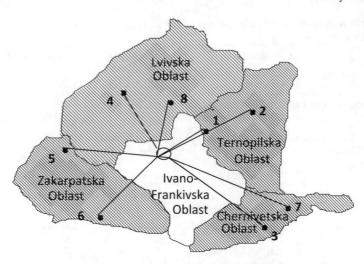

Fig. 8 Primary scheme of supply to the distribution center in the Ivano-Frankivsk region, 2018

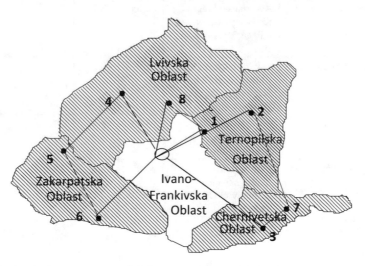

Fig. 9 Optimal scheme of transportation of vegetables to the distribution center in the Ivano-Frankivsk region, 2018

Thus, the deficit of vegetables in the regions: Kyiv region—130.5 Kt; Ivano-Frankivsk region—9.5 Kt, is proposed to be eliminated with the help of distribution centers to meet the demand of local consumers. The location coordinates of the distribution centers are determined by the center of gravity method, taking into account the required volumes of supply and the remoteness of neighboring regions that have an excess of production.

Due to the agreement between the suppliers and the distribution center, the transport task was solved using the Clark-Wright method, and the optimal way of transportation of vegetables to the distribution center is determined, taking into account the volumes of supplies, the location coordinates of the suppliers and the cargo capacity of the vehicle.

4 Conclusion

The national food market of Ukraine is heterogeneous in its structure, so each region has its own conditions for the vegetables production in particular and the functioning of the vegetable market in general.

The production of vegetable products in Ukraine is divided in the following climatic zones: Steppe, Forest-Steppe, Polissya and Carpathians, which are inherent in their peculiarities. Favorable conditions for growing cabbage and carrots are in the Polissya and Carpathian zones, in the Steppe—tomatoes, peppers, eggplants and onions, in the Forest-Steppe—beets of table and cucumbers.

In recent years, an increase in the consumption of vegetables has been observed, primarily due to the increase in their production, mainly in the private sector, rather than at the expense of production in large agricultural enterprises as the main source of vegetable supply to the trading network. Consumption of vegetables per capita also increased. To a certain extent, this is the result of a reduction in the population. A higher level of production is responsible for a higher level of consumption of vegetable products per capita. Of course, there is a surplus of consumption over production.

There is a positive tendency to increase production volumes, which is likely to continue in the near future. The reason for this is the dynamic development of the vegetable market in Ukraine in recent years, which, however, remains not saturated. Produced for consumption in fresh and processed form products will be delivered to the wholesale and retail trade, as well as the public catering network.

In order to increase the efficiency of the food market functioning of the region and to establish stable relations between commodity producers and consumers, the creation of a network of interregional wholesale food markets, which will promote the effective promotion of food products to the final consumer, should be considered as a priority.

According to the analysis of the possibility of satisfying the demand for vegetables in the region by regional producers, the shortage of these products in the Kyiv and Ivano-Frankivsk regions was revealed. Thus, it is these regions that need to create regional distribution centers to meet intra-regional consumption. The most rational way to transport costs is to transport products that lack the neighboring regions, apparently, under other conditions (transport tariffs, selling prices).

The most effective, according to the theoretical works on the management of the sales system, is the organization of sales of goods in accordance with the principles of distributive logistics. It covers the entire complex of tasks for the management of material flow at the site supplier—the consumer, from the moment of setting the implementation task and ending with the moment the product delivered from the scope of the supplier's attention. At the same time, the main share of the problem is the management of material flows, solved in the process of moving the already finished products to the consumer.

The direction of further research is to solve the problem of marketing support for enterprises of vegetable product subcomplex. It is possible on the basis of the formation of the appropriate infrastructure, the main units of which are marketing and marketing cooperatives, integrated into a vertical marketing system with regional wholesale points equipped with vegetable stores and regional (inter-district) vegetable exchanges. Each node is considered as an additional source of data that needs to be processed to formulate strategic decisions. The main coordinating links on the part of marketing support, in which it is advisable to concentrate data centers: interregional distribution center; regional (inter-district) exchange; marketing (sales) departments of business entities (agrarian enterprises, cooperatives).

References

1. Kolomiets K (2017) Improvement of the system of state regulation for providing the food security of the state. Market Infrastruct 12:49–54
2. Regional Logistics Competence Centers. The members of Open ENLoCC. http://openenlocc.net/files/open_enlocc_member_presentations_november_2015_lr.pdf, last accessed 02 Apr 2019
3. Kryvinska N (2012) Building consistent formal specification for the service enterprise agility foundation. J Serv Sci Res 4(2):235–269 (Springer, The Society of Service Science)
4. Kaczor S, Kryvinska N (2013) It is all about services—fundamentals, drivers, and business models. Journal of Service Science Research 5(2):125–154 (Springer, The Society of Service Science)
5. Allen T, Prosperi P (2016) Modeling sustainable food systems. Environ Manage 57:956–975. https://doi.org/10.1007/s00267-016-0664-8
6. Kirichenko L, Radivilova T, Zinkevich I (2017) Forecasting weakly correlated time series in tasks of electronic commerce. In: Proceedings of 2017 12th international scientific and technical conference on computer sciences and information technologies (CSIT), Lviv, pp 309–312. https://doi.org/10.1109/stc-csit.2017.8098793
7. Kirichenko L, Radivilova T, Zinkevich I (2018) Comparative analysis of conversion series forecasting in E-commerce tasks. In: Shakhovska N, Stepashko V (eds) Advances in intelligent systems and computing II. CSIT 2017. Advances in intelligent systems and computing, vol 689. Springer, Cham, pp 230–242. https://doi.org/10.1007/978-3-319-70581-1_16
8. Dobrynin I, Radivilova T, Maltseva N, Ageyev D (2018) Use of approaches to the methodology of factor analysis of information risks for the quantitative assessment of information risks based on the formation of cause-and-effect links. In: Proceedings of the 2018 international scientific-practical conference problems of infocommunications. Science and Technology (PIC S&T), Kharkiv, Ukraine, pp 229–232. https://doi.org/10.1109/infocommst.2018.8632022
9. Li X (2014) Operations management of logistics and supply chain: issues and directions. Discrete Dyn Nat Soc 2014:1–7. https://doi.org/10.1155/2014/701938
10. Molnar E, Molnar R, Kryvinska N, Gregus M (2014) Web intelligence in practice. J Serv Sci Res 6(1):149–172 (Springer, The Society of Service Science)

11. Kirichenko L, Bulakh V, Radivilova T (2017) Fractal time series analysis of social network activities. In: Proceedings of the 2017 4th international scientific-practical conference problems of infocommunications. Science and Technology (PIC S&T), Kharkov, Ukraine, pp 456–459. https://doi.org/10.1109/infocommst.2017.8246438
12. Kirichenko L, Radivilova T, Tkachenko A (2019) Comparative analysis of noisy time series clustering. In: Proceedings of the 3rd international conference on computational linguistics and intelligent systems (COLINS-2019). Volume I: Kharkiv, Ukraine, 18–19 April, pp 184–196
13. On the production and circulation of organic agricultural products and raw materials. http://www.zakon.rada.gov.ua/go/425-18, last accessed 2 Apr 2019
14. Ageyev D, Wehbe F (2013) Parametric synthesis of enterprise infocommunication systems using a multi-layer graph model. In: Proceeding of the IEEE 23rd international crimean conference microwave and telecommunication technology, pp 507–508
15. Agbo M, Rousselière D, Salanié J (2015) Agricultural marketing cooperatives with direct selling: A cooperative–non-cooperative game. J Econ Behav Organ 109:56–71. https://doi.org/10.1016/j.jebo.2014.11.003
16. Pysarenko V. Marketing of vegetable products (methodical and practical aspects): market of vegetable products: market conditions, subjects. https://agromage.com/stat_id.php?id=324, last accessed 2 Apr 2019
17. Pysarenko V. Marketing of vegetable products (methodical and practical aspects): marketing research of consumers, the retail and wholesale segment of the vegetable market. https://agromage.com/stat_id.php?id=325, last accessed 2 Apr 2019
18. The future of food and agriculture. http://www.fao.org/3/a-i6583e.pdf, last accessed 2 Apr 2019
19. Squares, gross collections and yields of crops, fruits, berries and grapes. Stat Bull. http://agroua.net/statistics, last accessed 2 Apr 2019
20. Kobylinska T (2018) Statistical evaluation of plant-growing branch. J Zhytomyr State Technol Univ Ser Econ 83(1):66–70. https://doi.org/10.26642/jen-2018-1(83)-66-70
21. Yu K. Vegetable market of open ground and greenhouse. http://agro-business.com.ua/agro/ekonomichnyi-hektar/item/10912-rynok-ovochiv-vidkrytoho-gruntu-ta-teplychnykh.html, last accessed 2 Apr 2019
22. Ageyev D (2010) NGN network planning according to criterion of provider's maximum profit. In: Proceeding of the IEEE international conference on modern problems of radio engineering, telecommunications and computer science (TCSET-2010), pp 256–256
23. Ageyev D, Al-Ansari A, Qasim N (2015) Multi-period LTE RAN and services planning for operator profit maximization. In: Proceeding of the IEEE the experience of designing and application of CAD systems in microelectronics, pp 25–27. https://doi.org/10.1109/cadsm.2015.7230786
24. Al-Dulaimi A, Al-Dulaimi M, Ageyev D (2016) Realization of resource blocks allocation in LTE downlink in the form of nonlinear optimization. In: Proceeding of the IEEE 13th international conference on modern problems of radio engineering, telecommunications and computer science (TCSET), pp 646–648. https://doi.org/10.1109/tcset.2016.7452140
25. Rodashchuk H, Solskyi O, Kutkovetska T (2018) Use of informational technologies in the logistics activities of agricultural enterprises. Sci Bull Polissia 2:175–182. https://doi.org/10.25140/2410-9576-2018-2-1(13)-175-182
26. Jeřábek K, Majercak P, Kliestik T, Valaskova K (2016) Clark Wright algoritam modela uštede koji se koristi kod rješavanja problema usmjeravanja u logistici opskrbe. Naše more. 63:115–119. https://doi.org/10.17818/NM/2016/SI7
27. Pysarenko V (2010) Prospects for the development of the vegetable industry. Economica APK 11:110–114

Optimization of the Process of Determining the Effectiveness of Advertising Communication

Bohdan Kovalskyi⬤, Tetyana Holubnyk⬤, Myroslava Dubnevych⬤,
Nadiia Pysanchyn⬤ and Zoryana Selmenska⬤

Abstract The advantages of all types of print advertising are listed. The process of perception of advertising by a potential consumer is described. Taking into account the model of the hierarchy of factors influencing the effectiveness of advertising communication, the numerical weights of factors are established on the basis of pairwise comparisons using the method of numerical consistency and a matrix of pairwise comparisons constructed on this basis. It uses the scale of the relative importance of objects by Saati. The optimization of weight values of factors is carried out and an optimized model of priority influence of factors influencing the effectiveness of advertising communication is synthesized. It is advisable to evaluate the validity of advertising messages using the PDI scale (Persuasive Discourse Inventory—assessment of the persuasiveness of messages) by TS Feltham, built on the Aristotle communication model: a convincing message based on trust in its source (ethos), emotionality message (pathos) and reasonability of arguments (logos). Advertising materials from Cambridge, San Francisco and Bristol universities and a student respondent were chosen for the study. It was designed using the method of multi-criteria optimization of alternative variants of the process of determining the effectiveness of advertising communication, based on the assignment and registration of certain indicators logos, ethos, pathos among the selected pre-university options. The choice of the optimal variant is determined by the utility functions. The calculation of alternative options is to use the Pareto principle, the essence of which is to choose among a multitude of indicators those that by their influence dominate over others and determine the optimal one.

Keywords Print advertising · Communication · Efficiency · Method · Optimization · Alternative option

B. Kovalskyi · T. Holubnyk (✉) · M. Dubnevych · N. Pysanchyn · Z. Selmenska
Ukrainian Academy of Printing, L'viv, Ukraine
e-mail: golubnyk@ukr.net

© Springer Nature Switzerland AG 2020
D. Ageyev et al. (eds.), *Data-Centric Business and Applications*,
Lecture Notes on Data Engineering and Communications Technologies 42,
https://doi.org/10.1007/978-3-030-35649-1_3

1   Introduction

Print advertising, which includes all types of print, including outdoor advertising, is the most important type of advertising. Evidence of this is both the history of advertising and advertising budgets. It was print advertising that served as a model for other types of advertising. Print advertising was, is and will be the main item of expenditure on advertising budgets.

Advertising images attract the attention of the consumer and express some key points of advertising. In addition, most advertising images are not able to cover the semantic space as a whole.

Special attention in the formation of an advertising product should be given to advertising text. Advertising text is a special genre of writing article material that aims to attract the attention of the target audience, interest it with a profitable offer, and convince to make a purchase or use the service. People working on advertising text are attracted by knowledge of text theory, through which the picture of the world shines through and reflects the cultural context beyond which a true understanding of the text is impossible.

Perhaps this is one of the most difficult genres. Because these texts can be neither good nor bad. They must be effective. Their main task is to turn an ordinary reader into a client. Advertising without text or with incorrectly written text does not work. On average, 5–7% of those who read the headline read the text. To keep people's attention and effectively sell them a product, they use tried and tested methods of writing advertising text.

Researchers of advertising text Rosenthal D. E. and Kohtev N. N. noted its uniqueness and thus formulated the specifics of the requirements for it: first, it must be specific and focused, advantageously distinguishing the advertised object from its own kind. Secondly, evidence-based, logically constructed, intelligible. Thirdly, brief, ignoring the secondary importance of the details and concise on the construction of phrases. Fourthly, original, unique in detail and at the same time corresponding generally accepted samples, as well as interesting, entertaining, witty. Fifth, he must inform and persuade. And, sixthly, it must be literary. Poor mastering of the original inevitably leads to quality losses in the translated text as well. Such is the minimum to which one should strive.

The perception of advertising for each person begins with a note of new information in the mind. The main task of any advertiser is to force the probable consumer to contact the advertising message and, therefore, influence the perception of advertising. However, the perception of advertising is limited to some factors, the most important of which may be a glut. Every person is exposed daily to a large number of advertising messages that compete with each other for the attention and perception of the consumer. For the correct perception by the consumer of your information, you should take into account certain features in your own marketing company. The whole process of perception of advertising by a potential consumer can be divided into several steps:

- Contact. The first and main step in the perception of advertising. High-quality contact ensures good placement of the advertising message. The advertising text or block must be placed so that it can be seen, heard, read as many potential consumers as possible.
- Awareness. After the first contact of the client with the advertising message, he becomes aware of the service or product. At this step comes the perception of advertising—the main motivating message. However, the main purpose of the advertiser is not the awareness of advertising, and awareness of the product that is promoted. Awareness is the first step in the perception of advertising.
- Awareness. The perception of advertising is based on simple handling. Awareness is an effort to understand the news. In itself, the perception of advertising relies on the active consciousness of the likely consumer. Any person first becomes interested, then finds out the details, later remembers the information received. Consciousness is extremely important for the perception of advertising, which contains a huge amount of information about the promoted product. Detailed information about the product, individuality, and properties of the product, competitive advantages are extremely necessary to understand. The perception of advertising in such a quantity of information simplifies clarity and ease of submission. Potential consumers will not understand the advertising message if the information in it is confusing and incomprehensible. It is necessary that a disinterested consumer simply and without difficulty understand the advertising message and evaluate the advertised product. The perception of advertising is easier if the information and proving properties are clear and accessible. When developing a marketing message, you need to provide features so that they are very simply understood by ordinary disinterested people.
- Persuasion. The perception of advertising is quite strongly tied to conviction. Advertising must prove to the person something. Quality advertising is required to maintain and change ideas, present facts and arguments, affect the emotional component and ideas of people. The perception of advertising is based on the trust in it. The furor of advertising depends entirely on the degree of trust in it. For each product and communication channel, there are ways to increase trust. The main thing is to understand that a positive perception of advertising without proper trust in it is unattainable.

Print advertising is primarily aimed at the visual perception of the information message chosen by the audience. It is outdoor advertising that has extremely high penetrating power.

Analysis of recent research and publications. Specialists in the theory of advertising describe and analyze it from different angles. Preferred advertising case with the disclosure of the main issues of Frank Jeffkins, Nazaykin A., Mudrov A. How to manage creative processes in the advertising business are presented Primak T. [1–3]. Characterized by the evolution of advertising in the works of Khavkina L., Bulakh T. [4]. However, detailed mathematical in-depth studies can optimize and uncover a new method for determining the effectiveness of advertising printed communication.

The aim of the work is to further optimize the model of the priority of the factors, determine the effectiveness of advertising using the PDI scale method of Feltham T. and calculate alternative process options.

2 Optimization of the Model of the Priority of the Factors on the Effectiveness of Advertising Communication

In a previous study, it was proved that the combination of the factors described is a set of $D = \{d_1, d_2, \ldots, d_9\}$ [5–10], from which you can define a subset $D_1 \in D$ most significant criteria.

Let's add the mathematical designation of indicators:

d_1—the thematic focus of the publication;

d_2—nature and features of the readership of the publication;

d_3—circulation;

d_4—territory of distribution of the publication;

d_5—periodicity of publication;

d_6—volume of the publication;

d_7—rating of the publication;

d_8—the cost of one advertising contact;

d_9—specific factors of a technical nature.

Based on the system analysis and mathematical modeling of hierarchies, the priority levels of factors are determined, which made it possible to construct a structural model of the hierarchy with an indication of the significance of their influence.

To further optimize the model of the prioritization of the factors, will allow at the beginning of the media planning of advertising communication to highlight the main factors in terms of their influence on the process.

For further research, it is necessary to establish the numerical weight of the factor on the basis of pairwise comparisons using the method of numerical consistency and a matrix of pairwise comparisons constructed on this basis [11–13]. It uses the scale of the relative importance of objects according to Saaty [11, 27] (Table 1).

Depending on the degree of mutual influence of factors, have in position $(k_1 k_2)$ value of matrix elements of pairwise comparisons A, tabulated (Table 1). Let him $A = (a_{ij})$. This matrix is inversely symmetric, i.e. $a_{ji} = 1/a_{ij}$, diagonal elements are equal to one. The lower part of the matrix is filled with inverse values. With minor differences between the weights of the factors, use intermediate values 2, 4, 6, 8.

When constructing the matrix of pairwise comparisons (Table 2), each element (factor) of the mnemonic column of names is compared with all elements of the first row of the table. The condition of the advantage is set according to the priority model (Fig. 1) and the scale estimates (Table 1).

The vector of priorities for the pairwise comparison matrix is obtained on the basis of the Saaty method. First we find the main eigenvector $D(d_1, d_2, \ldots, d_n)$ as the geometric mean of the elements of each row of the matrix, i.e.

Table 1 The scale of the relative importance of objects

Evaluation of importance	Comparison criteria	Explanation of the choice of criteria
1	Objects equivalent	No advantage k_1 above k_2
3	One object several another one prevails	There is a reason for having a weak advantages k_1 above k_2
5	One object another one prevails	There is a reason for having significant advantage k_1 above k_2
7	One object significantly another one prevails	There is a reason for having advantages k_1 above k_2
9	One object absolutely another one prevails	Absolute dominance k_1 above k_2 no doubt
2, 4, 6, 8	Intermediate values	Auxiliary comparative assessments

Table 2 Pairwise comparison matrix

	d_1	d_2	d_3	d_4	d_5	d_6	d_7	d_8	d_9
d_1	1	9	5	2	4	3	5	7	5
d_2	1/9	1	1/5	1/7	1/5	1/3	1/7	2	4
d_3	1/7	3	1	1/3	1/7	2	4	5	3
d_4	1/3	5	3	1	2	3	5	3	5
d_5	1/5	3	5	1/2	1	3	5	3	5
d_6	1/9	5	2	1/7	1/3	1	2	3	3
d_7	1/5	3	2	1/5	1/7	1/2	1	1/3	2
d_8	1/9	1/8	1/3	1/7	1/9	1/7	3	1	2
d_9	1/9	1/7	1/5	1/9	1/7	1/3	1/7	1/2	1

$$D_i = \sqrt[n]{a_{i1} \cdot a_{i2} \cdots \cdot a_{in}} \quad i = \overline{1, n} \tag{1}$$

As a result, we obtain the following vector:

$$D = (3.644;\ 0.396;\ 1.104;\ 2.357;\ 1.909;\ 0.994;\ 0.608;\ 0.341;\ 0.219)$$

Vector normalization D carried out according to the well-known formula:

$$D_{i\,norm} = \frac{\sqrt[n]{a_{i1} \cdot a_{i2} \cdots \cdot a_{in}} \quad i = \overline{1, n}}{\sum_{i=1}^{n} \sqrt[n]{a_{i1} \cdot a_{i2} \cdots \cdot a_{in}}} \tag{2}$$

Normalized Vector

$$D_{norm} = (0.314;\ 0.034;\ 0.095;\ 0.203;\ 0.164;\ 0.085;\ 0.052;\ 0.029;\ 0.018)$$

Fig. 1 Model of the
hierarchy of factors
influencing the effectiveness
of advertising
communication

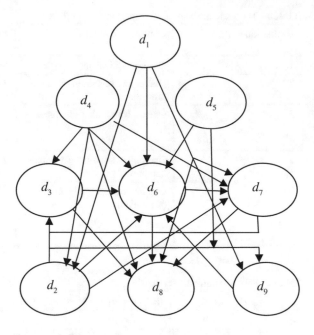

essentially determines the optimized numerical priorities of factors determining the effectiveness of advertising and establishes a preliminary formal result of solving the problem.

For a more convenient representation of the weight values of the factors, multiply the components of the vector D_{norm} by a factor for example, $k = 500$. We will get

$$D_{norm} \times k = (157;\ 17;\ 47.5;\ 101.5;\ 82;\ 42.5;\ 26;\ 14.5;\ 9).$$

We calculate the consistency estimate for the weighted values of the factors. To do this, multiply the matrix of pairwise comparisons on the right by the vector D_{norm} Get the normalized vector

$$D_{norm1} = (2.798;\ 0.321;\ 0.921;\ 1.799;\ 1.613;\ 0.817;\ 0.564;\ 0.355;\ 0.175).$$

Find the components of the eigenvector of the pairwise comparison matrix D_{norm2}. To do this, separate the components of the vector D_{norm1} on the corresponding components of the vector D_{HopM}. Get the vector

$$D_{norm2} = (8.888;\ 9.387;\ 9.652;\ 8.838;\ 9.782;\ 9.516;\ 10.73;\ 12.06;\ 9.216).$$

Maximum Eigenvalue λ_{max} positive inverse symmetric matrix A—this is the arithmetic mean of the vector D_{norm2}. After calculations, we have $\lambda_{max} = 9.787$. The decision score is determined by the consistency index.

Table 3 The random index value for matrices of the different order

Amount of objects	3	4	5	6	7	8	9	10	11	12	13	14
Reference value index	0.58	0.90	1.12	1.24	1.32	1.41	1.45	1.49	1.51	1.54	1.56	1.57

$$IU = \frac{\lambda\max - n}{n - 1}. \tag{3}$$

In our case, given the value $\lambda\max = 9.787$, and $n = 9$ get the value of the consistency index: $IU = 0.09$.

The value of the consistency index is compared with the reference values of the consistency index, the so-called random index (DI). The results are satisfactory if the index value does not exceed 10% of the reference value.

A table of random index values for matrices of a different order (which is equivalent to a different number of objects) is given in Table 3.

Comparing the calculated and reference index value for ten objects, and checking the inequality $IU < 0.1 \times DI$, we get: $0.09 < 0.1 \times 1.45$. The implementation of inequality confirms the adequacy of the solution of the problem.

The implementation of inequality confirms the adequacy of the solution of the problem $DU = IU / DI$ Insofar as $DI = 1.45$, that $DU = 0.06$. Results can be considered satisfactory if. So, we have a sufficient level of convergence of the process and proper consistency of expert judgments regarding pairwise comparisons of factors.

To obtain the weights of the factors, we proceed as follows. Considering the level of placement of factors, we assign them, starting from the lowest level, conditional numerical values reflecting the weight of the factor in the general scheme. In this low level, we assign a weight equal to 20 conventional units, and increase the weights of the factors of each next level by 20 conventional units. We get the next row of values: (d_9)—20; (d_8)—40; (d_2)—60; (d_7)—80; (d_6)—100; (d_3)—120; (d_5)—140; (d_4)—160; (d_1)—180.

Rearrange the obtained components according to the order of their placement in the matrix, that is: (d_1)—180; (d_2)—60; (d_3)—120; (d_4)—160; (d_5)—140; (d_6)—100; (d_7)—80; (d_8)—40, (d_9)—20. The numerical values of the factors are written in the form of components of some source vector

$$D_0 = (180; 60; 120; 160; 140; 100; 80; 40; 20).$$

Finally, the values of the factors obtained by different methods, we put for comparison in Table 4. For the purpose of further visualization, we use pro scaled values $D_n \times k$ (Figs. 2 and 3).

Since the values of the weights of the factors of the optimized vectors almost coincide, the components of any of them can serve as a basis for synthesizing an optimized

Table 4 Options for weight values of factors determine the effectiveness of advertising communication

i	1	2	3	4	5	6	7	8	9
D_0	180	60	120	160	140	100	80	40	20
D_n	0.314	0.034	0.095	0.203	0.164	0.085	0.052	0.029	0.018
$D_n \times k$	157	17	47.5	101.5	82	42.5	26	14.5	9

Fig. 2 Histogram of component weight values weekend (Do) and normalized vectors (Dn)

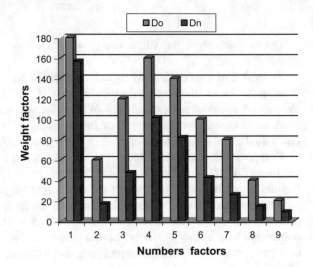

Fig. 3 A comparative graph of the weight values of the components of the source (Do) and normalized vector (Dn)

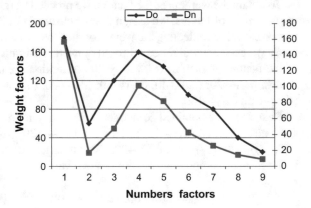

model of the influence of factors on the process of determining the effectiveness of advertising products (Fig. 4).

As a result of the study, a model of factors has been obtained, optimized by the priority of influence on the determination of the effectiveness of printed advertising

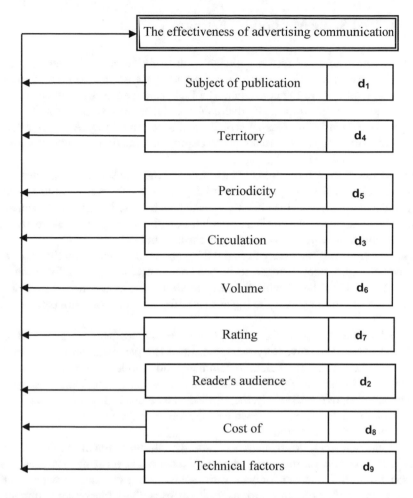

Fig. 4 Optimized model of factors influencing the effectiveness of advertising communication

products. The basis for the optimization of the model was the optimized weights of the factors.

A multi-level graphic model of Fig. 4, optimized by the method of pair comparison, compared with the model of Fig. 1, coincide in order of influence on the process, indicates the adequacy of the results obtained and the possibility of using any of these models as a baseline in further research.

3 The PDI Scale Method by Feltham T.

After the optimization of factors influencing the effectiveness of advertising communication, it is advisable to consider a method of measuring the effectiveness of
advertising messages to the consumer. The PDI scale method by Feltham T. is the
best way to assess the impact of advertising on consumers. Considering the emotional impact of marketing messages on the audience is a priority. After all, a person
in most life situations and in making choices in one direction or another is guided
by emotional intelligence [15].

Evaluation of the effectiveness of advertising is the sore spot of most enterprises.
Thousands of hryvnias are spent on advertising, and the effectiveness of these activities is often incomprehensible not only to managers but also to marketing specialists.
Assessing the increase in sales is a difficult task without analyzing the effectiveness
and profitability of advertising campaigns carried out [14].

That is why it was necessary to find the key moment of impact on the consumer.
At a certain point, people began to be considered not only as a rational but as a
spiritually developed being with different emotions. The assumption arose that it is
on the basis of emotions that the buyer makes the final decision and perceives the
advertisement.

It is advisable to evaluate the validity of advertising messages using the PDI scale
(Persuasive Discourse Inventory—assessment of the persuasiveness of messages)
by Feltham T., built on the Aristotle communication model: a convincing message
based on trust in its source (ethos), emotionality message (pathos) and reasonability
of arguments (logos). In the study, the respondent assesses 17 statements on a 7-point
scale, with the help of these assessments the value of the factors "ethos", "logos"
and "pathos" [14] is determined.

TS Feltham in 1994. He created a generalized, multidimensional, theoretically reasonable scale system for measuring the effectiveness of advertising messages. In his
opinion, using PDI, one can compare various advertisements, carry out quantitative
measurements and coordinate various ways of researching marketing communications. The essence of this scale is to measure the emotional reactions to advertising,
it shows that people feel after viewing an advertisement, and not what they learned
from it [14].

When using it, the respondent is given to mark the cells that best describe what
they see. If the roller completely corresponds to the characteristic at the beginning or
at the end of the scale, then tick the cell closest to the characteristic at the beginning
or at the end of the scale. If the roller does not correspond either to the characteristic
at the beginning of the scale, or to the characteristic at the end of the scale, then a
cell is marked in the middle [15] (Table 5).

The study is conducted according to the following algorithm [15]:

- View respondent promotional materials;
- The poll of the respondent according to the scale PDI;
- Calculation of the values of ethos, logos, pathos using formulas;

Table 5 Scale Feltham T. credibility scores advertising message

1	Senseless							Meaningful
2	Unreliable							Reliable
3	Stimulating							Not stimulating
4	Rational							Irrational
5	Energetic							Non-energy
6	Incredible							Probable
7	Refers to me							Doesn't appeal to me
8	Uninformative							Informative
9	Concerns me emotionally							Doesn't concern me emotionally
10	Logical							Illogical
11	Reliable							Unreliable
12	Leaves indifferent							Exciting
13	Trustworthy							Not credible
14	Touched my feelings							Didn't hurt my feelings
15	Believable							Implausible
16	Connected with facts							Not related to facts
17	Unspectacular							Exciting

- Graphic presentation of assessment results;
- Summary of the study (Tables 6, 7, 8 and 9).

Based on the derived formulas with TS Feltham, we will calculate:

Table 6 Ethos

Scale	№	Oppositions
E_1	6	Incredible–probable
E_2	15	Believable–implausible
E_3	2	Unreliable–reliable
E_4	11	Reliable–unreliable
E_5	13	Trustworthy–not credible

Table 7 Logos

Scale	№	Oppositions
L_1	4	Rational–irrational
L_2	8	Uninformative–informative
L_3	16	Connected with facts–not related to facts
L_4	1	Senseless–meaningful
L_5	10	Logical–illogical

Table 8 Pathos

Scale	№	Oppositions
P_1	14	Touched my feelings–didn't hurt my feelings
P_2	9	Concerns me emotionally–doesn't concern me emotionally
P_3	3	Stimulating–not stimulating
P_4	7	Refers to me–doesn't appeal to me
P_5	5	Energetic–non-energy
P_6	12	Leaves indifferent–exciting
P_7	17	Unspectacular–exciting

Table 9 Scale Feltham T. evaluation of the credibility of advertising messages

Cambridge University Campaigns

1	Senseless							✓		Meaningful
2	Unreliable							✓		Reliable
3	Stimulating				✓					Not stimulating
4	Rational	✓								Irrational
5	Energetic				✓					Non-energy
6	Incredible							✓		Probable
7	Refers to me				✓					Doesn't appeal to me
8	Uninformative							✓		Informative
9	Concerns me emotionally			✓						Doesn't concern me emotionally
10	Logical		✓							Illogical
11	Reliable	✓								Unreliable
12	Leaves indifferent					✓				Exciting
13	Trustworthy	✓								Not credible
14	Touched my feelings		✓							Didn't hurt my feelings
15	Believable			✓						Implausible
16	Connected with facts	✓								Not related to facts
17	Unspectacular				✓					Exciting

$$Logos = L_2 + (8 - L_1) + (8 - L_3) + L_4 + (8 - L_5) \tag{4}$$

$$Ethos = E_1 + (8 - E_2) + E_3 + (8 - E_4) + (8 - E_5) \tag{5}$$

$$Pathos = (8 - P_1) + (8 - P_2) + (8 - P_3)$$
$$+ P_4 + (8 - P_5) + (8 - P_6) + (8 - P_7) \tag{6}$$

Fig. 5 A histogram comparing the values of logo, ethos, pathos for promotional materials of the University of Cambridge

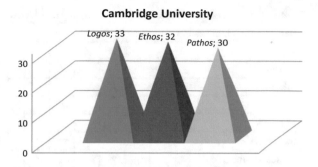

$$Logos = 33 \quad Ethos = 32 \quad Pathos = 30$$

Let's build a histogram (Fig. 5) comparing the values of the logo, ethos, pathos for promotional materials of the University of Cambridge.

Taking into account the histogram, which was obtained after assessing the persuasiveness of promotional materials of the University of Cambridge, one can say that the main emphasis was placed on the consistency and expediency of appeals to the target audience. Indeed, by logos (33 units) we understand the informative component of the message. The pathos is the least important (30 units), which speaks of the weak emotionality that the consumer should have when viewing advertising materials. Ethos—the perception of the reliability of the messages underlying advertising. Its value is 32, and this indicates sufficient confidence in the target audience.

Advertising agency Hub Strategy presented an advertising campaign for the University of San Francisco. The main idea of the action was humorous and extraordinary appeals to the audience. The background for the models was selected photos of the city in black and white. Hub Strategy wanted to move away from the typical San Francisco supply and the Golden Gate Bridge. On each layout used dark green color, the university logo and the main slogan of the university: "Change the world here" [16] (Table 10).

$$Logos = 19 \quad Ethos = 28 \quad Pathos = 27$$

Analyzing the histogram (Fig. 6), you can immediately notice the low value of logos (19), which indicates low consistency and insufficient reasonableness of advertisements. Although the value of ethos (28) and pathos (27) do not differ much from each other. In relation to the two components of the message of trust and emotion, there is a balance. However, this advertising campaign cannot be considered perfect through the low value of the logos.

An advertising campaign for the University of Bristol was created by Peloton Design along with Out of Home International and TMP Worldwide. The innovative approach will be in the use of a QR code, which is one of the types of matrix-type bar codes, which, when scanned from a mobile device, retains up to 7089 characters. For this promotion, the code was encrypted links to a unique page of

Table 10 Scale Feltham T. evaluating the credibility of advertising messages

San Francisco University Campaigns

Blindness			✓			Comprehended
Unreliable				✓		Reliable
Stimulating		✓				Not stimulating
Rational				✓		Irrational
Energetic	✓					Not energetic
Incredible					✓	Probable
Appeals to me					✓	Not appeals to me
Non-informative	✓					Informative
Touches me emotionally					✓	Does not touch me emotionally
Logical		✓				Illogical
Reliable	✓					Unreliable
Leaves indifferent					✓	Exciting
Deserves trust	✓					Not deserve trust
Hurt my feelings					✓	Did not hurt my feelings
Plausible		✓				Unlikely
Related to the facts	✓					Not related to the facts
Not addictive					✓	Exciting

Fig. 6 A histogram comparing the values of the logo, ethos, pathos for promotional materials of the University of San Francisco

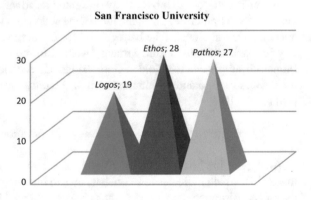

San Francisco University

the university. On it, visitors could get acquainted with additional information about the educational institution, additional places for training in the next year and other necessary information [17].

In addition, a special stylized design of the Clifton Bridge was developed, characterized by its simplicity. The slogan of the advertising campaign selected: «Learn more.» Which prompts you to use the QR code or open the page yourself in the absence of the necessary software on the mobile device [17, 18] (Table 11).

Table 11 Scale Faltham T. evaluating the credibility of advertising messages

Campaign of Bristol University								
Blindness						✓		Comprehended
Unreliable					✓			Reliable
Stimulating	✓							Not stimulating
Rational			✓					Irrational
Energetic		✓						Not energetic
Incredible					✓			Probable
Appeals to me	✓							Not appeals to me
Non-informative			✓					Informative
Touches me emotionally		✓						Does not touch me emotionally
Logical	✓							Illogical
Reliable	✓							Unreliable
Leaves indifferent						✓		Exciting
Deserves trust	✓							Not deserve trust
Hurt my feelings					✓			Did not hurt my feelings
Plausible		✓						Unlikely
Related to the facts					✓			Not related to the facts
Not addictive					✓			Exciting

$$Logos = 24 \quad Ethos = 31 \quad Pathos = 25$$

The construction of the histogram (Fig. 7) based on the calculations provides an opportunity to visually see the results of the study. One can observe a balance in the values of logos (24) and pathos (25), which indicates a uniform consistency and emotionality of the advertising message. However, there is a significant predominance of ethos (31)—confidence in billboards over the other components of Logos

Fig. 7 A histogram comparing the values of the logo, ethos, pathos for promotional materials of the University of Bristol

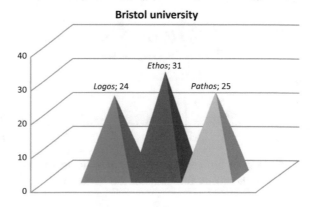

and Pathos. We cannot talk about the balance of this advertising campaign precisely because of the significant difference in the values of the evaluation of messages.

After completing all the calculations, you can talk about the effects of the advertising campaigns of universities. An indisputable advantage of the campaign of the University of Cambridge. Having the smallest difference in values for logos, ethos, and pathos. Also, high values among the three ranked universities.

Despite the histograms of the University of San Francisco and the University of Bristol, there is a noticeable lack of balance among the emotionality of the message of consistency and trust in it. Although all three campaigns can be considered successful, given the already received feedback from the target audience. However, creating your own advertising campaign should take advantage of the experience of the University of Cambridge.

4 Designing Alternative Options for Determining the Effectiveness of Advertising Communication

The next task concerns the design of alternative variants of the process of determining the effectiveness of advertising communication, based on the assignment and consideration of certain indicators logos, ethos, pathos in selected pre-university versions. The choice of the optimal variant is determined by the utility functions.

Further calculation of alternative options is to use the Pareto principle [19, 20], the essence of which is to choose among a variety of indicators of those who dominate over others in their effects. They form the so-called Pareto set $P(D)$, where $D \subset R^n$—many admissible solutions.

The task of multi-criteria optimization on the set D subject to the availability of target functions $f(x) = (f_1(x), \ldots, f_m(x))$ is to find the maximum value of the utility functions, that is $f_i(x) \to \max_{x \in D}, i = 1, m$ [21–25]. The multi-criteria choice of an alternative is based on the method of linear folding of criteria, the essence of which lies in the linear union of all particular objective functionals. $f_1, \ldots, f_m$ in one functional and selection of the maximum of them.

$$F(s, x) = \sum_{i=1}^{m} w_i f_i(x) \to \max_{x \in D}; \quad w \in W \tag{7}$$

where

$$W = \left\{ w = (w_1, \ldots, w_m)^{\mathrm{T}}; \quad w_i > 0; \quad \sum_{i=1}^{m} w_i = 1 \right\}.$$

Weights w_i indicate the priority of indicators and are identified with the numerical values of the utility functions of the indicators.

The utility functions of indicators or alternatives are determined based on the methods of multi-criteria utility theory [19, 26, 27]. Their essence is summarized as follows: if the criteria (in our study indicators) are independent in utility and advantage, then there is a utility function

$$U(x) = \sum_{i=1}^{m} w_i u_i(y_i) \tag{8}$$

where $U(x)$—multicriteria utility function $(0 \le U(x) \le 1)$ alternatives x; $u_i(y_i)$—utility function i-criteria $(0 \le u_i(y_i) \le 1)$; y_i—value of alternative x behind i criterion; w_i—libra i criteria, and $0 < w_i < 1$, $\sum_{i=1}^{m} w_i = 1$.

The practical implementation of theoretical calculations will begin with the formation of the Pareto set with mutually undominated indicators [19, 28–35], that is, those that have a predominant influence on the process of determining the effectiveness of advertising communication. They are presented in the method discussed earlier.

In an optimized multilevel model, three-dimensional space indices are mutually undiminished. For research it is enough to choose the following of them:

$$s_1 - Logos \quad s_2 - Ethos \quad s_3 - Pathos$$

In the general case, the implementation of each of the indicators in alternatives can be expressed as a degree of evaluation of advertising effectiveness (indicators: logos, ethos, pathos), the shares of which are given in percent. The sum of shares for one factor with the given options should be 100%.

Let the table define the possible values of the shares for each of the factors that meet the requirements for the elements of the Pareto set. For simplicity, set the multiplicity to 10% (Table 12).

With three alternative options for each of the indicators, such combinations of values are possible measures of its importance.

Perform a calculation of three options for determining the effectiveness of advertising communication, choosing for each of the indicators arbitrary combinations of Table 13.

Alternative variants of a particular institution will be denoted by A, B, C. Where, A—Cambridge University, B—San Francisco, C—University of Bristol. The relative evaluation of alternatives using the three-dimensional space of Aristotle and selected

Table 12 Shares of indicators for evaluating outdoor advertising

The name of the three-dimensional space	Degree of it importance in alternatives (in percent)									
Logos	10	20	30	40	50	60	70	80	90	100
Ethos	10	20	30	40	50	60	70	80	90	100
Pathos	10	20	30	40	50	60	70	80	90	100

Table 13 Combinations of values of shares of factors

Combinations of three-dimensional space values in percent			
10–10–80	20–10–70	30–10–60	40–10–50
10–20–70	20–20–60	30–20–50	40–20–40
10–30–60	20–30–50	30–30–40	40–30–30
10–40–50	20–40–40	30–40–30	40–40–20
10–50–40	20–50–30	30–50–20	40–50–10
10–60–30	20–60–20	30–60–10	50–10–40
10–70–20	20–70–10	60–10–30	50–20–30
10–80–10	70–10–20	60–20–20	50–30–20
80–10–10	70–20–10	60–30–10	50–20–10

Table 14 Evaluating alternatives for selected indicators

The name of the three-dimensional space	Libra indicators	Evaluating alternatives by indicators		
		A (%)	B (%)	C (%)
Logos	76 (s_1)	30	20	50
Ethos	91 (s_2)	40	20	40
Pathos	82 (s_3)	50	30	20

indicators and certain weights, which form the total value of the indicators of the three universities for outdoor advertising evaluation, is reflected in Table 14.

According to the above theorem, we have the following initial statements: u_{ij}—utility j alternatives ($j = 1, 2, 3$) behind i factor ($i = 1, \ldots, 4$); U_j—multi-variate utility evaluation j alternatives, and

$$U_j = \sum_{i=1}^{4} w_i u_{ij}; \quad j = 1, 2, 3 \tag{9}$$

where u_{ij}—utility j alternatives by i factor.

Considering that the names of three-dimensional space chosen for alternatives are practically new, we must first find their refined weights in accordance with the calculations made earlier, taking into account the output weights of Table 14. For this purpose, we apply the methodology used to analyze and evaluate the initial (full) set of indicators for determining the effectiveness of advertising. Using the Saaty method and scale, we will construct a matrix of pairwise comparisons obtained earlier and specified in the second column of Table 15.

Table 15 The matrix of pairwise comparisons of Pareto set factors

	Logos	Ethos	Pathos
Logos	1	1/5	1/3
Ethos	5	1	3
Pathos	3	1/3	1

After the matrix normalization procedures, we obtain the main eigenvector, the components of which are normalized without scaling are the desired refined values of the weights of the parameters:

$$w_1 = 0.104; \quad w_2 = 0.636; \quad w_3 = 0.258;$$

To find the utility functions of the factors, we construct pairwise comparison matrices that reflect the advantages of alternatives for each of the factors. The consistency of the results will be checked using the maximum eigenvalue of the vector of priorities λ_{max} for each of the matrices, the index of consistency IU and relation consistency DU, which limit values are given above.

For the formation of matrices on a scale of the relative importance of objects, we use Table 14—data estimates of alternatives for the selected indicators. Finally, we get the following results.

Logos	A	B	C
A	1	2	1/4
B	1/2	1	1/6
C	4	6	1

$$\lambda_{max} = 3.01; \quad IU = 0.01; \quad DU = 0.01.$$

The usefulness of alternatives in terms of the logos:

$$u_{11} = 0.192; \quad u_{12} = 0.106; \quad u_{13} = 0.7.$$

Ethos	A	B	C
A	1	2	1
B	1/2	1	1/2
C	1	2	1

$$\lambda_{max} = 3; \quad IU = 0; \quad DU = 0.$$

The usefulness of alternatives in terms of ethos:

$$u_{21} = 0.4; \quad u_{22} = 0.2; \quad u_{23} = 0.4.$$

Pathos	A	B	C
A	1	2	4
B	1/2	1	2
C	1/4	1/2	1

$$\lambda_{max} = 3; \quad IU = 0.00; \quad DU = 0.00.$$

The usefulness of alternatives in terms of pathos:

$$u_{31} = 0.571; \quad u_{32} = 0.285; \quad u_{33} = 0.142.$$

The values of the priority vector λ_{max} for each of the matrices, the index of consistency DU meet accepted requirements. Now, from Formula (8) we obtain multivariate estimates of alternatives using such a system of equalities:

$$U_1 = w_1 \cdot u_{11} + w_2 \cdot u_{21} + w_3 \cdot u_{31};$$
$$U_2 = w_1 \cdot u_{12} + w_2 \cdot u_{22} + w_3 \cdot u_{32};$$
$$U_3 = w_1 \cdot u_{13} + w_2 \cdot u_{23} + w_3 \cdot u_{33}. \tag{10}$$

We calculate the values of the utility functions for each of the three options for evaluating the effectiveness of advertising, taking into account (9):

$$U_1 = 0.104 \cdot 0.192 + 0.636 \cdot 0.4 + 0.258 \cdot 0.571 = 0.421686$$
$$U_2 = 0.104 \cdot 0.106 + 0.636 \cdot 0.2 + 0.258 \cdot 0.285 = 0.211754$$
$$U_3 = 0.104 \cdot 0.7 + 0.636 \cdot 0.4 + 0.258 \cdot 0.142 = 0.363836$$

According to the statement of the problem (7), a multi-criteria choice of an alternative is to determine the maximum value of the combined functional. As follows from the calculations, this condition corresponds to the option U_1. Thus, the advantage among the three given alternatives is provided by option A, which will be considered optimal from the point of view of evaluating the effectiveness of advertising.

5 Conclusion

On the basis of the constructed matrix of pairwise comparisons, we have optimized the weight values so far and synthesized an optimized model of the priority influence of factors influencing the effectiveness of advertising communication. Calculated rational method for assessing the impact of advertising on consumers by the method of PDI scales Feltham T. for three well-known foreign universities. The alternative variants of the process of determining the effectiveness of advertising communication are designed using the method of multi-criteria optimization. The criterion for choosing the optimal alternative is the maximum value of the unified functional.

As a result, the conducted studies provide an opportunity to optimize and calculate alternatives in other processes.

So, print advertising remains in demand. Given the selected audience, we can assume the effectiveness of the impact on the consumer. Properly formed advertising appeal attracts attention and thus penetrates and encourages action.

References

1. Jefkins F, Yadin D (2000) Advertising. Pearson Education Limited, Harlow
2. Dobrynin I, Radivilova T, Maltseva N, Ageyev D (2018) Use of approaches to the methodology of factor analysis of information risks for the quantitative assessment of information risks based on the formation of cause-and-effect links. In: Proceedings of the 2018 international scientific-practical conference problems of infocommunications. Science and Technology (PIC S&T), Kharkiv, Ukraine, pp 229–232. https://doi.org/10.1109/infocommst.2018.8632022
3. Primak T (2006) Advertising creative. KNEU, Kijiv
4. Khavkina L (2006) The basis of the advertising advertising activity: methodical materials for students of specialty "Journalism". View. 2, add. PE "Azamayev V.R." Kharkiv
5. Ogirko I, Dubnevych M, Holubnyk T, Pysanchyn N, Ogirko O (2018) Information technology to determine the effectiveness of advertising communication. IEEE international scientific-practical conference «Problems of Infocommunications. Science and Technology» (PIC S&T-2018), Kharkiv Ukraine, pp 752–756. https://doi.org/10.1109/infocommst.2018.8632016
6. Ageyev D, Wehbe F (2013) Parametric synthesis of enterprise infocommunication systems using a multi-layer graph model. In: Proceeding of the IEEE 23rd international crimean conference microwave & telecommunication technology, pp 507–508
7. Kirichenko L, Bulakh V, Radivilova T (2017) Fractal time series analysis of social network activities. In: Proceedings of the 2017 4th international scientific-practical conference problems of infocommunications. Science and Technology (PIC S&T), Kharkov, Ukraine, pp 456–459. https://doi.org/10.1109/infocommst.2017.8246438
8. Kirichenko L, Radivilova T, Tkachenko A (2019) Comparative analysis of noisy time series clustering. In: Proceedings of the 3rd international conference on computational linguistics and intelligent systems (COLINS-2019), vol I. Kharkiv, Ukraine, 18–19 Apr, pp 184–196
9. Kirichenko L, Radivilova T, Zinkevich I (2017) Forecasting weakly correlated time series in tasks of electronic commerce. In: Proceedings of 2017 12th international scientific and technical conference on computer sciences and information technologies (CSIT), Lviv, pp 309–312. https://doi.org/10.1109/stc-csit.2017.8098793

10. Kirichenko L, Radivilova T, Zinkevich I (2018) Comparative analysis of conversion series forecasting in E-commerce tasks. In: Shakhovska N, Stepashko V (eds) Advances in intelligent systems and computing II. CSIT 2017. Advances in intelligent systems and computing, vol 689. Springer, Cham, pp 230–242. https://doi.org/10.1007/978-3-319-70581-1_16
11. Saati T (1993) Decision making (hierarchy analysis method). Radio and communication
12. Zykov A (1987) Fundamentals of graph theory. The science, Moscow, Russia
13. Lesdon L (1990) Optimization of large systems. Nauka, Moscow
14. Kyshtymova I (2013) Image methodology and research methods. J Psychol Econ Manag 2:89–92
15. Kutlaliev A, Popov A (2006) Advertising effectiveness (professional editions for business). Exmo, Moscow
16. Writing a letter to the world, https://www.johnsonbanks.co.uk/work/dear-world-yours-cambridge. Last accessed 2019/03/21
17. Hub launches new brand campaign for USF, http://www.thesfegotist.com/news/local/2012/april/4/hub-launches-new-brand-campaign-usf. Last accessed 2019/03/15
18. University of Bristol billboard campaign features QR Codes to attract new students, http://www.thedrum.com/news/2011/11/28/university-bristol-billboard-campaign-features-qr-codes-attract-new-students. Last accessed 2019/03/15
19. Bartish M (2009) Operations research. Part 3. Decision making and game theory. Published by Center of Ivan Franko LNU, Lviv, Ukraine
20. Makarov I, Vinogradskaya T (1982) Theory of choice and decision making—the science, Moscow, Russia
21. Lamets V (2004) Sistemny analize. Introductory course. KNURE, Kharkiv
22. Samarskyi A (2005) Mathematical modeling: ideas. Methods. PHYSMATLIT, Moscow
23. Mesarovic M (1966) Foundations of the general theory of systems. World, Moscow
24. Mesarovic M (1978) General systems theory: mathematical foundations. World, Moscow
25. Mesarovic M (1973) Theory of hierarchical multilevel systems. World, Moscow
26. Larichev O (2003) Theory and methods of decision making. Logos, Moscow
27. Syavavko M (2007) Information system "Nechitky expert" Published by Center of Ivan Franko LNU, Lviv, Ukraine
28. Kryvinska N (2012) Building consistent formal specification for the service enterprise agility foundation. J Serv Sci Res 4(2):235–269
29. Kaczor S, Kryvinska N (2013) It is all about services—fundamentals, drivers, and business models. J Serv Sci Res 5(2):125–154
30. Gregus M, Kryvinska N (2015) Service orientation of enterprises—aspects, dimensions, technologies. Comenius University in Bratislava
31. Kryvinska N, Gregus M (2014) SOA and its business value in requirements, features, practices and methodologies. Comenius University in Bratislava
32. Molnar E, Molnar R, Kryvinska N, Gregus M (2014) Web intelligence in practice. J Serv Sci Res 6(1):149–172
33. Ageyev D (2010) NGN network planning according to criterion of provider's maximum profit. In: Proceeding of the IEEE international conference on modern problems of radio engineering, telecommunications and computer science (TCSET-2010), pp 256–256
34. Ageyev D, Al-Ansari A, Qasim N (2015) Multi-period LTE RAN and services planning for operator profit maximization. In: Proceeding of the IEEE the experience of designing and application of CAD systems in microelectronics, pp 25–27. https://doi.org/10.1109/cadsm.2015.7230786
35. Al-Dulaimi A, Al-Dulaimi M, Ageyev D (2016) Realization of resource blocks allocation in LTE downlink in the form of nonlinear optimization. In: Proceeding of the IEEE 13th international conference on modern problems of radio engineering, telecommunications and computer science (TCSET), pp 646–648. https://doi.org/10.1109/tcset.2016.7452140

Using Recurrent Neural Networks for Data-Centric Business

Serhii Leoshchenko[ID]**, Andrii Oliinyk**[ID]**, Sergey Subbotin**[ID]
and Tetiana Zaiko[ID]

Abstract Over the last decade, print and online publications have published a huge number of articles with proposals for the introduction and use of neural networks in existing business and research. And just two years ago, many leading IT companies showed the world already created smart applications in the field of neural networks, which indicates the uniqueness and relevance of this technology. This is easily explained by the fact that systems based on neural networks are able to perform complex business tasks more efficiently and cheaper than the people. While working with big data, the probability of error remains relatively low. Unlike humans, neural networks are more stable. With long-term high loads, the efficiency of solving problems by the neural network does not sag. Finally, neural networks make free the people from monotonous computing operations and enable creative implementation. However, the issue of choosing the most appropriate topology and type of neural networks remains extremely relevant. The best results are demonstrated by recurrent neural networks. This paper is devoted to the review and proposing on the use of recurrent neural networks in data-centric business. The tests and analysis of their results demonstrated the relevance of the use of modern architectures of artificial neural networks.

Keywords Artificial neural networks · Recurrent neural networks · Training · LSTM · GRU · BRNN · Backpropagation through the time · Data-centric business

S. Leoshchenko (✉) · A. Oliinyk · S. Subbotin · T. Zaiko
Department of Software Tools, National University "Zaporizhzhia Polytechnic", Zaporizhzhia 69063, Ukraine
e-mail: sedrikleo@gmail.com

A. Oliinyk
e-mail: olejnikaa@gmail.com

S. Subbotin
e-mail: subbotin@zntu.edu.ua

T. Zaiko
e-mail: nika270202@gmail.com

© Springer Nature Switzerland AG 2020
D. Ageyev et al. (eds.), *Data-Centric Business and Applications*,
Lecture Notes on Data Engineering and Communications Technologies 42,
https://doi.org/10.1007/978-3-030-35649-1_4

1   Introduction

No matter how much times is constructed buyer persona, segmentation is very averaged. The human brain cannot process huge amounts of data, draw millions of conclusions and scenarios, remember them and effectively apply them. But it can be done by the machines, by its trainings and creating the artificial neural networks (ANN) for business.

Neural networks are one of the ways to perception of sensory information by artificial or machine intelligence. It is obvious that biological neural networks have become the prototype of neural networks. That is, human ways of obtaining visual information, which is two-thirds of the total sensory traffic [1].

ANN are a fundamentally new approach to solving problems that were solved by algorithmic programming. There is no need to build algorithms for all possible cases of development of the system or processes in the system and strive to anticipate all the options and describe the logic for them [2–6]. Neural networks allow on the basis of a large number of accumulated data to independently find patterns and relationships in the previously not explicit aspects and use this information for further forecasting, classification and management of data and processes.

Primitive, it can be roughly described the process of neural networks: the neural network at the input receives the big data. Then this data is analyzed, the neural network is trained (machine learning) on the basis of positive and negative examples. In the process of training, the structure of the neural network is formed, which in the future can solve the problems of identification, classification, prediction [7].

2   Implementation ANN to Data-Centric Business

ANN will help businesses to promote merchandise and services. If marketers now rely on average segmentation and targeting, in the near future ANN, knowing so much about the user and being able to process information very quickly, will predict exactly what the person wants. A deep understanding of the desires and needs of the consumer is the key to success. The ability to see problems, behavior and life cycle in dynamics gives business great opportunities [8–20].

This is how some services, such as Google Play Music [21], already work. The neural network examines the user day by day and finally gives recommendations that exactly match the interests. ANN do not stops on it: their recommendations anticipate interests. The person comes to listen to one group, but in the end happy to be stuck on the songs that are listened to in 15 years that caused a big bout of nostalgia.

Neural networks will help companies avoid technology commoditization. This is situation when all products, services and goods are rapidly becoming the same. Appearance, quality and benefit—all the same as competitors. How to attract customers? Will help a special approach to each. And this will make neural networks [22–33].

For example, a customer buys a new TV. The employee of the store standard recommends to buy supplies: remote, cable, mounting system. The ANN in a matter of seconds will analyze the customer and his purchase and offer to buy a coffee machine. The ANN "knows" that the client just furnish the kitchen and think about buying a good coffee machine [34–41].

It is advisable to use ANNs in business tasks, in such situations [42–47]:

- accumulated a huge amount of different data;
- there are no working methods for processing and systematization of these data yet;
- data is corrupted, corrupted, incomplete, or not systematized;
- there is a wide variety of data and at first glance it is difficult to establish links and patterns between them.

Possibilities and examples of possible applications of neural networks and machine learning for business tasks [48]:

- predictions, forecasting, risk assessment (forecasting of demand, sales volume, average bill, frequency of sales, loading of equipment to optimize the amount of cash, warehouses and other resources);
- searching of the trends, correlations, tendencies. Prediction of further development of the system and prediction of possible changes. Artificial intelligence has significantly improved the mechanisms of recommendations in online stores and services. Algorithms based on machine learning analyze clients behavior on the site and compare it with millions of other users. All in order to determine which product you will buy with the highest probability:
- recognition of photo, video, audio content. Various services and online applications using recognition technology;
- machine learning for dialogue by computer systems. To automate the activities of operators in online chats, telephone operators and messengers. The development of chat-bots. ANN that analyze natural language can be used to create chatbots that allow customers to get the necessary information about the company's products. This will reduce costs for call center teams [49].

3 Recurrent Neural Networks for Data-Centric Business

Recurrent neural networks (RNN) are a type of ANN intended to recognize certain patterns in a sequence of data, regardless of their kind: text, images, audio streams (including spoken words), genome or numerical sequences coming from sensors. The main difference between RNN and other architectures is the so-called presence of memory. RNN retains the previous value in his condition. For clarity, consider an example [50, 51].

In the process of life people do not start to think every second "from scratch". That is, the erasing of all previously accumulated knowledge does not occur and any mental activity is based on existing knowledge and experience. All our knowledge and thoughts are permanent.

Traditional ANN do not have this property, and this is their main drawback. RNN, on the contrary, retain the previous values, which in our example can be regarded as an analogue of human knowledge and experience [50–56].

It can be said that the RNN builds dynamic models, that is, models that change over time in such a way that it is possible to achieve sufficient accuracy, depending on the context of the examples that have been provided.

Let's imagine a fragment of RNN (Fig. 1). The neural network fragment A takes the input value of x_t and returns the value of ht. The presence of feedback allows you to transfer information from one step of the network to another.

The complexity and incomprehensibility of the RNN is the presence of feedbacks [50, 57]. Therefore, for greater clarity, it is possible to present the RNN as several copies of the same network, each of which transmits information to the subsequent copy [58–64]. The scheme with expanded feedback is shown in Fig. 2.

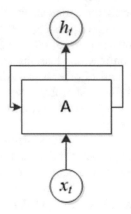

Fig. 1 A fragment of RNN

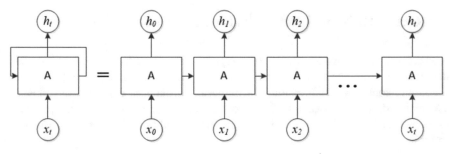

Fig. 2 The scheme with extensive feedback connections

A detailed representation of RNN in the chain testifies to the close ties with strings and lists. That is why many agree that RNN is the most natural architecture of neural networks for working with this type of data.

However, during the use of RNN often face the problem of "long-term dependencies" [65, 66].

To perform most of the primitive—current tasks need only recent information. In this case, it can be argued that the distance between the actual information and the place where it is needed is small (Fig. 3).

But in real practice, such cases are rare. In speech recognition experiments, it was noted that a large context is needed to predict the end of phrases. And in such cases, the gap between the actual information and the point of its application can become very large, and with the growth of this distance, the RNN lose the ability to link information (Fig. 4).

In theory, problems with processing long-term dependencies in RNN can be solved by careful selection or artificial change of the problem to be solved. Unfortunately, in practice it is too resource-intensive. This problem is investigated in detail Hochreiter (1991) [67] and Bengio with co-authors (1994) [68, 69]; they found undeniable reasons why it can be difficult.

Over the past few years, RNN has been used with incredible success for a number of tasks: speech recognition, language modeling, translation, image recognition, etc.

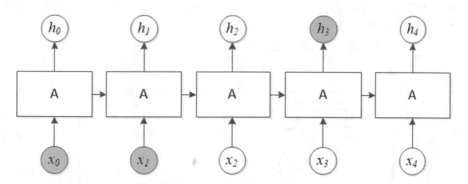

Fig. 3 RNN can train to use information "from the past"

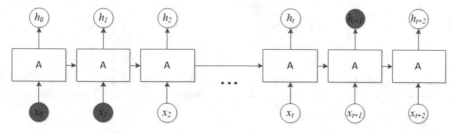

Fig. 4 RNN lose the ability to communicate information

A significant role in these successes belongs to LSTM—an unusual modification of the recurrent neural network, which in many tasks is much superior to the standard version. Almost all the impressive results of RNN are achieved with the help of LSTM. LSTM are designed specifically to avoid the problem of long-term dependence. Storing information for long periods of time is their normal behavior, not something they struggle to learn.

4 Using the LSTM Architecture

Long short-term memory (LSTM) is a special kind of architecture of RNN, capable of learning long—term dependencies. Submitted by Hochreiter and Schmidhuber in 1997 [70], and then improved and adapted popular in the works of many other researchers. They perfectly solve a number of various tasks and are now widely used.

Compared to an ordinary schematic diagram of RNN (Fig. 1) and LSTM networks (Fig. 5) it is possible to note a more complex representation of LSTM due to the presence of additional elements called gates that need to control the data flows. Depending on its state, the gate may or may not miss a signal.

LSTM consists of the following parts (Fig. 5).

- network input (input);
- network output (output);

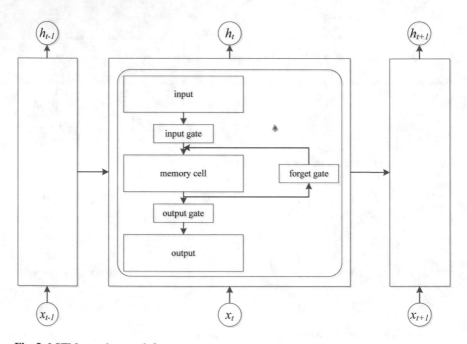

Fig. 5 LSTM neural network fragment

- memory or network status (memory cell);
- block to clear the memory (forget gate);
- memory update unit (input gate);
- output gate.

The operation of the network is as follows [71]:

1. Set the initial state of the network output $y = 0$, the state of memory $C = 0$.
2. Submit another signal x from the sequence of inputs X to the network input.
3. Calculate the state of the input gate (g_i)

$$g_i = \sigma(W_i x + R_i y + P_i C + b_i), \tag{1}$$

where x is input, W_i is the input weight of network for input gate, y is previous network output, R_i is the output weight of network for input gate, C is state of the memory cell, P_i is the memory cell weight of network for input gate, b_i is the shift for the value input gate, σ is value activation function input gate.

4. Calculate the state of the forget gate (g_f)

$$g_f = \sigma(W_f x + R_f y + P_f C + b_f), \tag{2}$$

where x is input, W_f is the input weight of network for forget gate, y is previous network output, R_f is the output weight of network for forget gate, C is state of the memory cell, P_f is the memory cell weight of network forget gate, b_f is the shift for the value forget gate, σ is value activation function forget gate.

5. Calculate the forget memory (Z)

$$Z = \tanh(W_m x + R_m y + b_m), \tag{3}$$

where x is input, W_m is the weight of the input network to forget memory, y is previous network output, R_m is weight network output to forget memory, b_m is the shift for the value of forget memory, tanh is value activation function forget memory.

6. Update the state of memory cell (C)

$$C(t + 1) = Z \times g_i + C \times (t)g_f, \tag{4}$$

where $C(t)$ is network memory state at a time t, Z is forget memory, g_i is value of input gate, g_f is value of forget gate. $\times$ is operation element-by-element multiplication.

7. Calculate the state of the output gate (g_o)

$$g_o = \sigma(W_o x + R_o y + P_o C + b_o), \tag{5}$$

where x is input, W_o is the input weight of network for output gate, y is previous network output, R_o is the output weight of network for output gate, C is state of the memory cell, P_o is the memory cell weight of network for output gate, b_o is the shift for the value output gate, σ is value activation function output gate.

8. Calculate the state of the output (y)

$$y = \tanh(C) \times g_o, \tag{6}$$

where C is state of the memory cell, g_o is value of output gate, × is operation element-by-element multiplication.

9. If the queue of the input sequence X is empty, then the end of the work; otherwise, the transition to 2 stage.

As can be seen from the description of the network LSTM really better cope with the tasks due to the complicated architecture. However, such an architecture requires additional calculations and, as a consequence, additional resources for calculations and data storage. Moreover, researches have shown that before LSTM networks training, data should be filtered, that is, the informative features selection should be made, and it also requires additional computing power.

In 2014, K. Cho introduced the Gated Recurrent Unit (GRU) model based on the same principles as LSTM, but uses fewer filters and operations for computation [72].

5   GRU Architecture and Its Peculiarit

GRU is a small variation of the previous network. This has one less filter, and links are implemented differently. The update gate determines how much information will remain from the previous state and how much will be taken from the previous layer. The reset gate works like a forget gate. Schematic representation of both types of networks demonstrates the differences (Fig. 6) [72–74].

In the case of comparison of LSTM and GRU, preference is given to GRU in the cases:

1. when it is necessary to speed up the work time;
2. when it is impossible to select informative features or they are not enough;

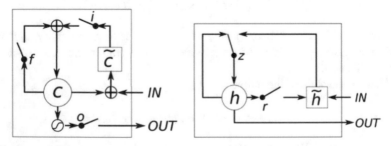

Fig. 6 Comparison of LSTM and GRU

3. when memory resources are limited;
4. in speech recognition problems.

6 Bidirectional Recurrent Neural Networks

Bidirectional recurrent neural networks (BRNNs) are based on the idea that the output at time t can depend not only on the previous elements in the sequence, but also on the future. For example, if you want to predict the missing word in a sequence, given both the left and right context. BRNNs are quite simple. It's just two RNN, stacked on top of each other. The output is then calculated based on the latent state of both RNNS [74].

The difference is that these networks use not only data from the "past", but also from the "future". For example, a typical LSTM network trained to guess the word "fish" for feeding the letters one at a time, and bidirectional for feeding yet, and the next letter of the sequence. Such networks are capable, for example, not only to expand the image at the edges, but also to fill the holes inside [75, 76].

7 The Implementation of Data-Centric Business Using Recurrent Neural Networks

The state of any element of the data-centric business is characterized by a large number of state parameters (features), the values of which can be obtained the results of observations of customers or market [77, 78]. ANN allow to classify and predict further trends, new desires and needs of customers, global changes in the market.

After training, the network is able to predict the future value of a sequence based on several previous values or some existing factors. It should be noted that prediction is possible only when the previous changes do to some extent determine the future.

On ANN, the problem of prediction is formalized through the problem of pattern recognition. Data on the predicted variable for a certain period of time form an image, the class of which is determined by the value of the predicted variable at some point in time outside the given period, i.e. the value of the variable through the prediction interval.

The general scheme of diagnosis and prediction the state of the technical system element with the use of ANN is shown in Figs. 7 and 8.

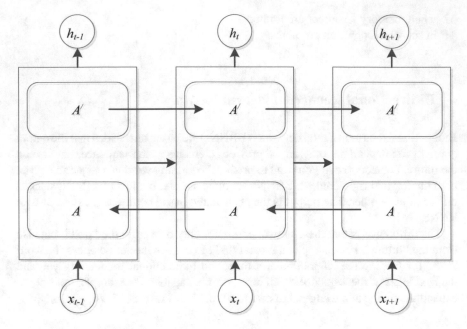

Fig. 7 Structure overview of the BRNN

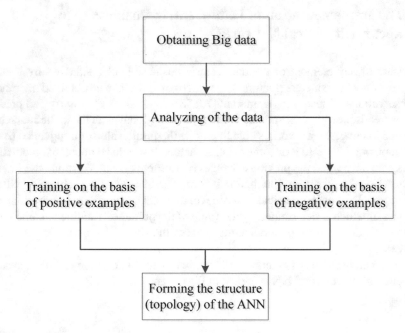

Fig. 8 General scheme of functioning ANN in data-centric business applications

8 Testing and Experimental Comparison of RNN

As a model (building tool) it is supposed to use LSTM, GRU and BRNN networks. The test will identify the potential for each architecture. For the experiment, it is going to use two test samples, which are freely available [79, 80]. One of the test samples went through the feature selection process [81], specifically to compare the work of two architectures on different data. As a training method for neural networks was used Backpropagation through time [82]. Information about both test samples is presented in Table 1.

During the testing, special attention was paid to the main parameters of the work: spent time, the average value of the error and error threshold values.

Table 2 shows the main test results. Tables 3 and 4 shows the distribution of the number of iterations necessary for the synthesis of RNN.

Figure 9 shows a diagram of the distribution of iterations between testing network architectures during experiments on two samples. Based on the diagram, you can clearly see the difference in iterations between the two experiments. This increase in iterations is explained by a significant increase in the number of information objects in the sample Online Retail Data Set. However, it is worth noting a fairly similar picture of the overall performance for networks. That is, the architecture of

Table 1 Characteristics of data samples

Criterion	Characteristic	Criterion	Characteristic
Dow Jones index data set			
Data set characteristics	Multivariate	Number of instances	750
Attribute characteristics	Integer, real	Number of attributes	16
Online retail data set			
Data set characteristics	Multivariate	Number of instances	541,909
Attribute characteristics	Integer, real	Number of attributes	8

Table 2 Results of the testing

Criterion	Time	E_{min}	E_{max}	E
Dow Jones index data set				
LSTM	10634.08	0.006	0.011	0.009
GRU	5139.69	0.018	0.026	0.022
BRNN	19125.30	0.003	0.009	0.006
Online retail data set				
LSTM	24525.65	0.009	0.015	0.012
GRU	12354.36	0.014	0.032	0.023
BRNN	30262.52	0.008	0.013	0.011

Table 3 The table of distribution of number of iterations

Type of ANN	Number of iterations
Dow Jones index data set	
LSTM	104
GRU	52
BRNN	185

Table 4 The table of distribution of number of iterations

Type of ANN	Number of iterations
Online retail data set	
LSTM	486
GRU	382
BRNN	676

Fig. 9 The diagram of distribution of number of iterations

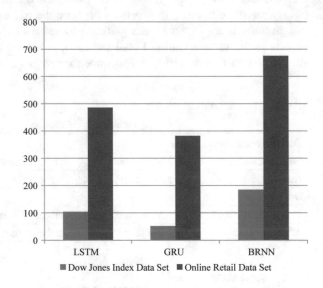

GRU requires much less iterations and, accordingly, their execution time, but in the case of BRNN requires much more computing operations and, accordingly, time to perform them.

For Figs. 10 and 11 shows the distribution of time spent between iterations. At the Fig. 10 presents the graph of time allocation with sample Dow Jones Index Data Set.

At the Fig. 11 presents the graph of time allocation with sample Online Retail Data Set.

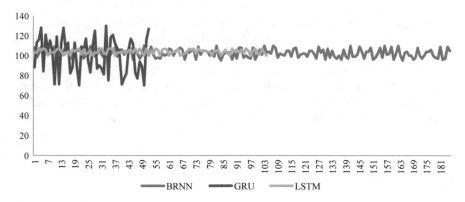

Fig. 10 Time allocation with sample Dow Jones index data set

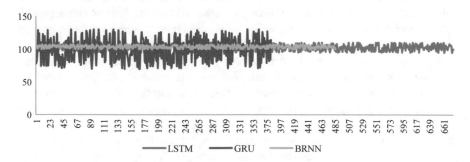

Fig. 11 Time allocation with sample online retail data set

9   Analysis of the Results

From the results can be drawn the following conclusions. Firstly, in both cases, the
time spent on LSTM and BRNN is greater than on GRU. However, in the case of
LSTM and BRNN increases proportionally with the increase of the input data. For
GRU, the time change is quite abrupt.

Secondly, it should be noted that the size of the average error for LSTM or BRNN
networks is smaller in both cases, what characterizes their work as more accurate.
When working with the technical diagnosis of complex systems, which may depend
on human life, this parameter is the most important. Moreover, it is important to note
that when testing a GRU architecture with input data that has not been processed,
the gap between the minimum error value and the maximum is too large.

Third, for when working with ANN, the problem of overfitting is quite common.
With the increase in the number of information features, the prediction accuracy often
falls. Especially if there are many uninformative features (little correlating with the
target variable) in the data. Therefore, the feature selection process, a stage in the
work with the ANN, which must not be missed in any case, especially given the

recent impressive results of new methods and approaches [83, 84]. Therefore, using preprocessed samples as input data will significantly reduce the time required for further processing.

Special attention should be paid to comparing the results of LSTM and BRNN. In testing, it was noted that LSTM accuracy results are inferior to BRN accuracy. However, the time for LSTM training is much shorter. From this it follows that the most optimal would be the use of LSTM networks.

10   Conclusion

ANN are indispensable for data analysis, in particular for preliminary analysis or selection, for identifying "outliers" or gross errors of the decision-maker. It is advisable to use neural network methods in problems with incomplete or uninformative data, especially in problems where the solution can be found intuitively, and the traditional mathematical models do not give the desired result.

Methods of neural networks can be used independently or serve as an excellent addition to traditional methods of statistical analysis, most of which are associated with the construction of models based on certain assumptions and theoretical conclusions (for example, that the desired dependence is linear or that some variable has a normal distribution). The neural network approach is not related to such assumptions—it is equally suitable for linear and complex nonlinear relationships, but is especially effective in exploratory data analysis, when the goal is to find out whether there are dependencies between variables. In this case, the data may be incomplete, contradictory and even deliberately distorted. If there is some connection between the input and output data, even if it is not detected by traditional correlation methods, the neural network is able to automatically tune to it with a given degree of accuracy. In addition, modern neural networks have additional capabilities: they allow to evaluate the comparative importance of different types of input information, reduce its volume without losing significant data, recognize the symptoms of approaching critical situations, etc.

The results of research and their analysis show that in the future for data-centric business as a model (building tool) it is better to use RNN, exactly LSTM network architecture. The architecture of GRU has proved to be a good alternative, but due to its innovation in some cases it is inferior in quality of work. In case of using BRNN architecture, special attention should be paid to optimizing the process of network training (synthesis). One of the solutions, for example, can be the use of neuroevolutionary [85–87] approaches that have proven themselves well when working with RNN. Such approaches will help to significantly reduce the training time of the network, while maintaining the accuracy of forecasting.

Perhaps in the future, when more research is done and GRU architecture requirements become more stringent, as they were once with LSTM, it will be a better solution.

It is also important to pay attention to new approaches for training NN that can facilitate and speed up the work. Thus, the process of technical diagnostics of infocommunication systems will be much easier.

Acknowledgements The work was performed as part of the project "Methods and means of decision-making for data processing in intellectual recognition systems" (number of state registration 0117U003920) of Zaporizhzhia National Technical University.

References

1. Tigreat P (2017) Sparsity, redundancy and robustness in artificial neural networks for learning and memory. Ecole nationale supérieure Mines-Télécom, Atlantique
2. Miller B, Ranum D (2013) Problem solving with algorithms and data structures, 3rd edn. Beedle & Associates, Franklin
3. Kryvinska N (2012) Building consistent formal specification for the service enterprise agility foundation. J Serv Sci Res 4(2):235–269
4. Kaczor S, Kryvinska N (2013) It is all about services—fundamentals, drivers, and business models. J Serv Sci Res 5(2):125–154
5. Gregus M, Kryvinska N (2015) Service orientation of enterprises—aspects, dimensions, technologies. Comenius University in Bratislava
6. Kryvinska N, Gregus M (2014) SOA and its business value in requirements, features, practices and methodologies. Comenius University in Bratislava
7. Montavon G, Samek W, Müller K-R (2017) Methods for interpreting and understanding deep neural networks. In: 2017 IEEE international conference on acoustics, speech and signal processing (ICASSP), pp 1–14
8. Xu S, Chen L (2008) A novel approach for determining the optimal number of hidden layer neurons for FNN's and its application in data mining. In: Proceeding 5th international conference on information technology and applications (ICITA 2008), pp 683–685
9. Fulcher J, Zhang M, Xu S (2006) Application of higher-order neural networks to financial time-series prediction. Artificial neural networks in finance and manufacturing, pp 80–108. https://doi.org/10.4018/978-1-59140-670-9.ch005
10. Smith KA, Gupta Jatinder JND (2000) Neural networks in business: techniques and applications for the operations researcher. Comput Oper Res 27(11–12):1023–1044. https://doi.org/10.1016/s0305-0548(99)00141-0
11. Tkáč M, Verner R (2016) Artificial neural networks in business: two decades of research. Appl Soft Comput 38:788–804. https://doi.org/10.1016/j.asoc.2015.09.040
12. Zhang GP (2004) Neural networks in business forecasting. Idea Group Publishing
13. Nisbet R, Elder J, Miner G (2009) Handbook of statistical analysis and data mining applications. Academic Press
14. Tufféry S (2011) Data mining and statistics for decision making. Wiley
15. Abbas O (2005) Neural networks in business forecasting. Int J Comput (IJC) 114–128
16. Lisboa PJG, Vellido A, Edisbury B (2000) Business applications of neural networks. In: Progress in neural processing, vol 13. https://doi.org/10.1142/4238
17. Limitations of neural networks grow clearer in business. https://searchenterpriseai.techtarget.com/feature/Limitations-of-neural-networks-grow-clearer-in-business
18. Neural networks for beginners: popular types and applications. https://blog.statsbot.co/neural-networks-for-beginners-d99f2235efca
19. Mitchell D, Pavur R (2002) Using modular neural networks for business decisions. Manag Decis 40(1):58–63. https://doi.org/10.1108/00251740210413361

20. Worku GB, Malik AK, Rao A (2018) Predicting business distress using neural network in sme-arab region. Int Rev Adv Bus Manag Law 1(1):68–84. https://doi.org/10.30585/irabml.v1i1.68
21. Google play music. https://play.google.com/music/listen
22. Companies using machine learning in cool ways. https://www.wordstream.com/blog/ws/2017/07/28/machine-learning-applications
23. Tarsauliya A, Kala R, Tiwari R, Shukla A (2011) Financial time series forecast using neural network ensembles. In: Hybrid particle swarm optimization with biased mutation applied to load flow computation in electrical power systems, pp 480–488. https://doi.org/10.1007/978-3-642-21515-5_57
24. Rudenko O, Bezsonov O, Romanyk O (2019) Neural network time series prediction based on multilayer perceptron. Dev Manag 17(1):23–34. https://doi.org/10.21511/dm.5(1).2019.03
25. Kharusi SAl, Murthy SRY (2017) Financial sustainability of private higher education institutions: the case of publicly traded educational institutions. Invest Manag Financ Innov 14(3):25–38. https://doi.org/10.21511/imfi.14(3).2017.03
26. Ghosh B, Kozarević E (2018) Identifying explosive behavioral trace in the CNX nifty index: a quantum finance approach. Invest Manag Financ Innov 15(1):208–223. https://doi.org/10.21511/imfi.15(1).2018.18
27. Abiodun OI, Jantan A, Omolara AE, Dada KV, Mohamed NA, Arshad H (2018) State-of-the-art in artificial neural network applications: a survey. Heliyon 4(11):1–41. https://doi.org/10.1016/j.heliyon.2018.e00938
28. Lee KS, Ahn KH (2019) Artificial neural network analysis of spontaneous preterm labor and birth and its major determinants. J Korean Med Sci 34(16):1–10. https://doi.org/10.3346/jkms.2019.34.e128
29. Walczak S (2012) Methodological triangulation using neural networks for business research. Adv Artif Neural Syst 2012, Article ID 517234. https://doi.org/10.1155/2012/517234
30. Samsudin R, Shabri A, Saad P (2010) A comparison of time series forecasting using support vector machine and artificial neural network model. J Appl Sci 10:950–958. https://doi.org/10.3923/jas.2010.950.958
31. Pavlyshenko B (2019) Machine-learning models for sales time series forecasting. In: Proceedings of the 2018 IEEE second international conference on data stream mining & processing (DSMP), Lviv, Ukraine, pp 1–11. https://doi.org/10.3390/data4010015
32. Pai PF, Lin CS (2005) A hybrid ARIMA and support vector machines model in stock price forecasting. Omega 33:505–597
33. Pelikan E, de Groot C, Wurtz D (1992) Power consumption in West-Bohemia: improved forecasts with decorrelating connectionist networks. Neural Netw World 2:701–712
34. Chakraborty K, Mehrotra K, Mohan CK, Ranka S (1992) Forecasting the behavior of multivariate time series using neural networks. Neural Netw 5:961–970
35. Chen A, Leung MT, Hazem D (2003) Application of neural networks to an emerging financial market: forecasting and trading the Taiwan Stock Index. Comput Oper Res 30:901–923
36. Chen KY, Wang CH (2007) A hybrid SARIMA and support vector machines in forecasting the production values of the machinery industry in Taiwan. Expert Syst Appl 32:54–264
37. Ginzburg I, Horn D (1994) Combined neural networks for time series analysis. Adv Neural Inf Process Syst 6:224–231
38. Oliinyk A, Zaiko T, Subbotin S (2014) Training sample reduction based on association rules for neuro-fuzzy networks synthesis. Opt Mem Neural Netw 23(2):89–95. https://doi.org/10.3103/S1060992X14020039
39. Oliinyk A, Oliinyk O, Subbotin S (2012) Agent technologies for feature selection. Cybern Syst Anal 48(2):257–267. https://doi.org/10.1007/s10559-012-9405-z
40. Oliinyk A, Subbotin S (2015) The decision tree construction based on a stochastic search for the neuro-fuzzy network synthesis. Opt Mem Neural Netw 24(1):18–27. https://doi.org/10.3103/S1060992X15010038

41. Oliinyk A, Fedorchenko I, Stepanenko A, Rud M, Goncharenko D (2018) Evolutionary method for solving the traveling salesman problem. In: Proceedings of the 2018 international scientific-practical conference problems of infocommunications. Science and technology (PIC S&T), Kharkiv, Ukraine, pp 331–339. https://doi.org/10.1109/infocommst.2018.8632033

42. Molnár E, Molnár R, Kryvinska N, Greguš M (2014) Web intelligence in practice. J Serv Sci Res 6(1):149–172

43. Kirichenko L, Radivilova T, Zinkevich I (2017) Forecasting weakly correlated time series in tasks of electronic commerce. In: Proceedings of 2017 12th international scientific and technical conference on computer sciences and information technologies (CSIT), Lviv, pp 309–312. https://doi.org/10.1109/stc-csit.2017.8098793

44. Kirichenko L, Radivilova T, Zinkevich I (2018) Comparative analysis of conversion series forecasting in e-commerce tasks. In: Shakhovska N, Stepashko V (eds) Advances in intelligent systems and computing II, CSIT 2017. Advances in intelligent systems and computing, vol 689. Springer, Cham, pp 230–242. https://doi.org/10.1007/978-3-319-70581-1_16

45. Dobrynin I, Radivilova T, Maltseva N, Ageyev, D (2018) Use of approaches to the methodology of factor analysis of information risks for the quantitative assessment of information risks based on the formation of cause-and-effect links. In: Proceedings of the 2018 international scientific-practical conference problems of infocommunications. Science and technology (PIC S&T), Kharkiv, Ukraine, pp 229–232. https://doi.org/10.1109/infocommst.2018.8632022

46. Kirichenko L, Bulakh V, Radivilova T (2017) Fractal time series analysis of social network activities. In: Proceedings of the 2017 4th international scientific-practical conference problems of infocommunications. Science and technology (PIC S&T), Kharkov, Ukraine, pp 456–459. https://doi.org/10.1109/infocommst.2017.8246438

47. Kirichenko L, Radivilova T, Tkachenko A (2019) Comparative analysis of noisy time series clustering. In: Proceedings of the 3rd international conference on computational linguistics and intelligent systems (COLINS-2019), vol I, Kharkiv, Ukraine, 18–19 Apr, pp 184–196

48. Zhang G, Patuwo BE, Hu MY (1998) Forecasting with artificial neural networks: the state of the art. Int J Forecast 14:35–62

49. Boiano S, Borda A, Gaia G, Rossi S, Cuomo P (2018) Chatbots and new audience opportunities for museums and heritage organisations. In: Electronic visualisation and the arts, pp 164–171, London, UK

50. Leoshchenko S, Oliinyk A., Subbotin, S, Zaiko T (2018) Using modern architectures of recurrent neural networks for technical diagnosis of complex systems. In: Proceedings of the 2018 international scientific-practical conference problems of infocommunications. science and technology (PIC S&T), Kharkiv, Ukraine, pp 411–416. https://doi.org/10.1109/infocommst.2018.8632015

51. Korbicz J, Kowa M (2007) Neuro-fuzzy networks and their application to fault detection of dynamical systems. Eng Appl Artif Intell Arch 20(5):609–617

52. Katsuba Y, Grigorieva L (2016) Application of artificial neural networks in vehicles' design self-diagnostic systems for safety reasons. In: 12th international conference "organization and traffic safety management in large cities". St. Petersburg, SPbOTSIC-2016, pp 283–287

53. Ageyev D (2010) NGN network planning according to criterion of provider's maximum profit. In: Proceeding of the IEEE international conference on modern problems of radio engineering, telecommunications and computer science (TCSET-2010), pp 256–256

54. Ageyev D, Al-Ansari A, Qasim N (2015) Multi-Period LTE RAN and services planning for operator profit maximization. In: Proceeding of the IEEE the experience of designing and application of CAD systems in microelectronics, pp 25–27. https://doi.org/10.1109/cadsm.2015.7230786

55. Ageyev D, Wehbe F (2013) Parametric synthesis of enterprise infocommunication systems using a multi-layer graph model. In: Proceeding of the IEEE 23rd international crimean conference microwave & telecommunication technology, pp 507–508

56. Al-Dulaimi A, Al-Dulaimi M, Ageyev D (2016) Realization of resource blocks allocation in lte downlink in the form of nonlinear optimization. In: Proceeding of the IEEE 13th international conference on modern problems of radio engineering, telecommunications and computer science (TCSET), pp 646–648. https://doi.org/10.1109/tcset.2016.7452140

57. Patan K (2008) Artificial neural networks for the modelling and fault diagnosis of technical processes. Springer, Berlin and Heidelberg GmbH & Co. K, India
58. Shkarupylo V, Skrupsky S, Oliinyk A, Kolpakova T (2017) Development of stratified approach to software defined networks simulation. East Eur J Enterp Technol 89(5/9):67–73. https://doi.org/10.15587/1729-4061.2017.110142
59. Kolpakova, T., Oliinyk, A., Lovkin, V (2017) Improved method of group decision making in expert systems based on competitive agents selection. In: Proceedings of the IEEE first Ukraine conference on electrical and computer engineering (UKRCON), Institute of Electrical and Electronics Engineers, pp 939–943. https://doi.org/10.1109/ukrcon.2017.8100388
60. Subbotin, S., Oliinyk, A., Skrupsky, S (2015) Individual prediction of the hypertensive patient condition based on computational intelligence. In: Proceedings of the international conference on information and digital technologies (IDT 2015), 25 Aug 2015, pp 348–356. https://doi.org/10.1109/dt.2015.7222996
61. Alsayaydeh JA, Shkarupylo V, Hamid MS, Skrupsky S, Oliinyk A (2018) Stratified model of the internet of things infrastructure. J Eng Appl Sci 13(20):8634–8638. https://doi.org/10.3923/jeasci.2018.8634.8638
62. Limarev I, Subbotin S, Oliinyk A, Drokin I (2019) Diagnostic signal nonstationarity reduction to predict the helicopter transmission state on the basis of intelligent information technologies. In: Proceedings of the second international workshop on computer modeling and intelligent systems (CMIS-2019), dblp key: conf/cmis/LimarevSOD19, pp 510–522
63. Stepanenko A, Oliinyk A, Fedorchenko I, Kuzmin V, Kuzmina M, Goncharenko D (2019) Analysis of echo-pulse images of layered structures. the method of signal under space. In: Proceedings of the second international workshop on computer modeling and intelligent systems (CMIS-2019), dblp key: conf/cmis/StepanenkoOFKKG19, pp 755–770
64. Fedorchenko I, Oliinyk A, Stepanenko A, Zaiko T, Svyrydenko A, Goncharenko D (2019) Genetic method of image processing for motor vehicle recognition. In: Proceedings of the second international workshop on computer modeling and intelligent systems (CMIS-2019), dblp key: conf/cmis/FedorchenkoOSZS19, pp 211–226
65. Lipton ZC, Kale DC, Elkan C, Wetzel RC (2016) Learning to diagnose with LSTM recurrent neural networks. In: International conference on learning representations (ICLR 2016), pp 1–18
66. Greff K, Kumar Srivastava, R, Koutník J, Steunebrink BR, Schmidhuber J (2015) LSTM: a search space odyssey, arXiv preprint arXiv:1503.04069
67. Hochreiter S (1991) Untersuchungen zu dynamischen neuronalen, Netzen. Diploma thesis, TU Munich
68. Bengio Y, Simard P, Frasconi P (1994) Long-term dependencies with gradient descent is difficult. IEEE Trans Neural Netw 5(2):157–166
69. Bengio Y, Frasconi P (1995) Diffusion of credit in markovian models. In: Advances in neural information processing systems (NIPS'1994), vol 7. MIT Press, Cambridge, pp 553–560
70. Hochreiter S, Schmidhuber J (1997) Long short-term memory. Neural Comput 9(8):1735–1780
71. LSTM recurrent network. http://mechanoid.kiev.ua/neural-net-lstm.html
72. Cho K, Merrienboer B, Gülçehre Ç, Bahdanau D, Bougares F, Schwenk H, Bengio Y (2014) Learning phrase representations using RNN encoder-decoder for statistical machine translation. In: Moschitti A, Pang B, Daelemans W (eds) EMNLP, pp 1724–1734
73. Chung J, Gulcehre C, Cho K, Bengio Y (2014) Empirical evaluation of gated recurrent neural networks on sequence, modeling. In: Comment: presented in NIPS 2014 deep learning and representation learning workshop, arXiv preprint arXiv:1412.3555
74. Budyilskiy DV (2015) GRU and LSTM: modern recurrent neural networks. Molodoy uchenyiy, vol 15, pp 51–54
75. Berglund M, Raiko T, Honkala M, Kärkkäinen L, Vetek A, Karhunen J (2015) Bidirectional recurrent neural networks as generative models. Reconstructing gaps in time series. NIPS
76. Lipton ZC, Berkowitz J, Elkan C (2015) A critical review of recurrent neural networks for sequence learning. In: International conference on learning representations, pp 1–32

77. Yarymbash D, Kotsur M, Subbotin S, Oliinyk A (2017) A new simulation approach of the electromagnetic fields in electrical machines. In: Information and digital technologies: international conference IDT'2017, Zilina, 5–7 July 2017: proceedings of the conference, Zilina: Institute of Electrical and Electronics Engineers, pp 429–434. https://doi.org/10.1109/dt.2017. 8024332

78. Stepanenko O, Oliinyk A, Deineha L, Zaiko T (2018) Development of the method for decomposition of superpositions of unknown pulsed signals using the second order adaptive spectral analysis. East Eur J Enterp Technol 92(2/9):48–54. https://doi.org/10.15587/1729-4061.2018. 126578

79. Dow Jones index data set. https://archive.ics.uci.edu/ml/datasets/dow+jones+index

80. Online retail data set. https://archive.ics.uci.edu/ml/datasets/online+retail

81. Oliinyk A, Subbotin, S, Lovkin V, Leoshchenko S, Zaiko T (2018) Feature selection based on parallel stochastic computing. In: Proceedings of the IEEE 13th international scientific and technical conference on computer sciences and information technologies (CSIT'2018), pp 347–351. https://doi.org/10.1109/stc-csit.2018.8526729

82. Werbos P (1990) Backpropagation through time: what it does and how to do it. Proc IEEE 78(10):1550–1560. https://doi.org/10.1109/5.58337

83. Oliinyk A, Subbotin S, Lovkin V, Leoshchenko S, Zaiko T (2018) Development of the indicator set of the features informativeness estimation for recognition and diagnostic model synthesis. In: Proceedings of the 14th international conference on advanced trends in radioelectronics, telecommunications and computer engineering (TCSET 2018), pp 903–908. https://doi.org/ 10.1109/tcset.2018.8336342

84. Oliinyk A, Leoshchenko S, Lovkin V, Subbotin S, Zaiko T (2018) Parallel data reduction method for complex technical objects and processes. In: Proceedings of the 9th international conference on dependable systems, services and technologies (DESSERT'2018), IEEE Catalog number: CFP18P47-ART 978-1-5386-5903-8, pp 526–532. https://doi.org/10.1109/dessert.2018. 8409184

85. Leoshchenko S, Oliinyk A, Subbotin S, Gorobii N, Zaiko T (2018) Synthesis of artificial neural networks using a modified genetic algorithm. In: Proceedings of the the of the 1st international workshop on informatics & data-driven medicine (IDDM 2018), dblp key: conf/iddm/PerovaBSKR18, pp 1–13

86. Leoshchenko S, Oliinyk A, Skrupsky S, Subbotin, S, Lytvyn V (2019) Parallel genetic method for the synthesis of recurrent neural networks for using in medicine. In: Proceedings of the second international workshop on computer modeling and intelligent systems (CMIS-2019), dblp key: conf/cmis/LeoshchenkoOSSL19, pp 1–17

87. Leoshchenko S, Oliinyk A, Skrupsky S, Subbotin S, Gorobii N, Shkarupylo V (2019) Modification of the genetic method for neuroevolution synthesis of neural network models for medical diagnosis. In: Proceedings of the second international workshop on computer modeling and intelligent systems (CMIS-2019), dblp key: conf/cmis/LeoshchenkoOSGS19, pp 143–158

System Model for Decision Making Support in Logistics and Quality Management Business Processes of Manufacture Enterprise

Yuliia Leshchenko and **Alina Yelizieva**

Abstract This paper is dedicated to research in logistics tasks decision making support area. The goal of the research is minimalizing of time and cost for decision making support in logistics business processes of manufacture enterprise taking according quality management methods. The subject of the research is logistics business processes in manufacture enterprise management. For obtaining the results, systems analysis methods were used to develop a procurement system management model; algebra of languages and algebra of algorithms for describing the sequence of solving local procurement management problems, methods of full-factor experiment for estimation of environment influence to logistic chain components. Business processes in manufacture enterprise management taking into account quality control are represented. The system model of decision making support including logistics processes and quality management tasks is offered. These model components are subscribed by algebra of algorithms equations taking into account business processes as stations and processes results as transfer conditions. The considered model takes into account methods of quality management in conditions of required resources, risks of success execution and deadline.

Keywords Logistic business processes · Quality management methods · Automate models · Algebra of algorithm · Algebra of relationships · Model of environment evaluation · Full factor experiment

1 Introduction

Significant difficulties in producing new types of products may be due to the lack of quality materials and advanced technical resources or difficulties in their delivery [1–3]. Therefore, timely solution of logistics tasks of supplying new materials,

Y. Leshchenko (✉) · A. Yelizieva
National Aerospace University "Kharkiv Aviation Institute", 17, Chkalov Str., Kharkiv, Ukraine
e-mail: j.leshhenko@khai.edu

A. Yelizieva
e-mail: a.elizeva@khai.edu

© Springer Nature Switzerland AG 2020　　　　　　　　　　　　　　　　　93
D. Ageyev et al. (eds.), *Data-Centric Business and Applications*,
Lecture Notes on Data Engineering and Communications Technologies 42,
https://doi.org/10.1007/978-3-030-35649-1_5

equipment, determining the required volume and range of resources, the possibilities of existing supplies and the selection of new suppliers is an important problem.

Nowadays, the tasks of managing the logistic processes of production, resource procurement planning and implementation of transportation services are quite worked out [4–7]. However, there remain unresolved issues related to the assessment of possible changes in the external environment, the composition of new material and technical resources and the planning of the supply process at the different stages of the manufacture enterprise product life cycle.

There are enough methods to control the quality of production. Providing modern quality requirements is the main area of aerospace products production to enhance its competitiveness. Most of enterprises can increase the level of product quality step by step, having regard to their limited economic possibilities [8, 9]. Therefore, the actual theme of this study, which addresses and solves the problem of improving the quality of aerospace products through the use of a strategy of post-graduate improvement of quality, taking into account the logistic chain production process.

1.1 Development of Procurement and Quality Management Models Based on Discrete Converter

In the conditions of the transition to the market model of the economy, enterprises have the opportunity to independently plan their main directions and conditions of their activity, determine the organizational forms of management, choose the type of economic activity, manage the labor, material, financial and information resources, choose their business partners.

Nowadays industrial enterprises function in disadvantageous conditions. Adaptation of business entities to a market economy occurs slowly, with many difficulties (imperfection of economic policy, tax system, unfavorable conditions of the investment climate, etc.). In recent years, enterprises have experienced a decrease in profits, an increase in the share of costs due to excessive formation of material stocks, a lack of reliable and timely information about the position on the market, and competitors.

The enterprise development management system is designed to ensure the integrity of the analysis and coordination of all processes and the optimization of technological (production) options, organizational, financial and economic aspects of the enterprise. Planning and assessing the feasibility of enterprise development plans is based on the articulated global strategic goal and coordination of the sub-goals of each of the incoming subsystems.

Thus, to increase the profitability of the enterprise, it is necessary [10–14]:

- to produce high-quality products, systematically update them and provide services in accordance with the demand and existing production capabilities;
- rational use of production resources, taking into account their interchangeability;
- to develop a strategy and tactics of enterprise behavior and to correct them in accordance with changing circumstances;

- systematically introduce everything new and advanced in production, in the organization of labor and management;
- to ensure the competitiveness of the enterprise and products, to maintain a high image of the enterprise.

That's why for increase of enterprise effectiveness it's necessary to develop model based on system approach taking into account influence of different factors:

- components of environment;
- logistics chain;
- enterprise parameters of functioning.

The high-tech production can be represented in the form of a logistic chain, elements of which are consistently linked (by inputs and outputs) among themselves (Fig. 1). At the upper level, such a representation corresponds to the logistics chain— "supply—production—sales" [15, 16].

According to Fig. 1 the sequences of actions at the life cycle stages are formalized in the processes models. There are four main components that describe the actions to solve the problems of procurement management: the factor of change in the state of the external environment, the quality management tasks, the parameter of the enterprise's activity and the problem of procurement management.

Logistics system operates in the context of constant changes related to the market of suppliers, economic conditions, competitors, constantly changing technologies, employees, goals, costs and customers. The external environment is divided into [10–12]:

- microenvironment is an environment of direct influence on the enterprise, which is created by suppliers of material and technical resources, consumers of products

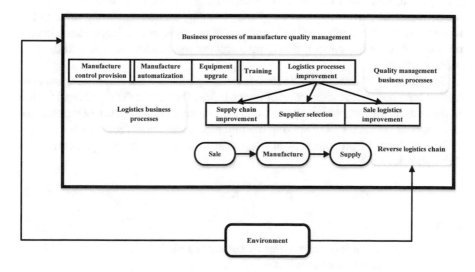

Fig. 1 Structure of business processes system model

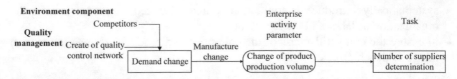

Fig. 2 Business processes at the life cycle stage of "growth"

(services) of an enterprise, trade and marketing intermediaries, competitors, government agencies, financial institutions, insurance companies and other contact audiences;
- macro environment affecting the enterprise and its microenvironment. It includes the natural, demographic, scientific and technical, economic, ecological, political and international environment.

The business processes sequence of decision making taking into account quality of logistics tasks and environment components is considered [13]. At the life cycle stage of "growth" business processes are represented at the Fig. 2.

The "growth" stage of life cycle products is accompanied by an increase in the sales volume of the manufactured products, which makes it possible to increase production volumes in order to maximize profit. According to the results of marketing research at the stage of pre-design research, at the "growth" stage there is an increase in the volume of demand for manufactured products, which leads to a change in the volume of production and, consequently, the volume of resources required for production.

For a formalized description of the sequence of business processes, an apparatus of automaton models was chosen. The equations of the processes in the algebra of languages determine the set of admissible values and describe the changes in the states of all elements of a discrete system, which makes it possible to determine the sequence of performing tasks and transforming the corresponding information in procurement management. The equations in the algebra of relations describe the possible states of the system, taking into account the conditions of transitions and control actions.

The presence of control implies two interacting blocks—the controlling and controlled components, the interaction of which is described using the regular expression of the relation algebra, which allows to take into account the influence of the external environment on the result of solving the purchase management problems.

The problem of procurement management in general form is represented as a finite state automate:

$$A = (X, f, a_0, F), \tag{1}$$

where

X	set of input characters or automate alphabet;
f	function of transitions;
a_0	set of begin automate states;

Fig. 3 Automate model of decision making process at the "growth" stage

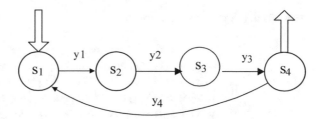

$\{a_0, a_1, \ldots, a_n\}$ set of automate states;

F set of end automate states.

The sequence of business processes at the stage of "growth" is represented by automate model, in which input signal is information about competitors (sale volume, quality production etc.) and output signal is number of potential suppliers for production provision of necessary material and technical resources (Fig. 3) [15, 16].

The internal signal is information about competitors (such as sales, product quality, etc.), and the output signal is the number of potential suppliers to supply the production with the necessary material and technical resources.

As model states can be used parameters:

- changes of environment parameters;
- task of quality management;
- changes of manufacture parameters.

The model states are:

- s_1—demand growth;
- s_2—quality management tasks;
- s_3—change of production volume;
- s_4—contracting with suppliers.

The transitions functions are:

- y_1—quality management methods implementation;
- y_2—determination of product production volume;
- y_3—task solution of supplier's number determination;
- y_4—possibility of repetition in condition of environment change.

Process equations in algebra of languages have the form:

$$F_1 = s_1 \vee F_4 s_1;$$
$$F_2 = s_2 \vee F_1 s_2;$$
$$F_3 = s_3 \vee F_2 s_3;$$
$$F_4 = s_4 \vee F_3 s_4. \tag{2}$$

Transform the process equations:

$$
\begin{aligned}
F_1 &= s_1 \lor s_1 s_4 \lor F_3 s_1 s_4 = s_1 \lor s_1 s_4 \lor s_1 s_3 s_4 \lor F_2 s_1 s_3 s_4 \\
    &= s_1 \lor s_1 s_4 \lor s_1 s_3 s_4 \lor s_1 s_2 s_3 s_4 \lor (s_1 s_2 s_3 s_4)^*; \\
F_2 &= s_2 \lor s_1 s_2 \lor F_4 s_1 s_2 = s_2 \lor s_1 s_2 \lor s_1 s_2 s_4 \lor F_3 s_1 s_2 s_4 \\
    &= s_2 \lor s_1 s_2 \lor s_1 s_2 s_4 \lor s_1 s_2 s_3 s_4 \lor (s_1 s_2 s_3 s_4)^*; \\
F_3 &= s_3 \lor s_2 s_3 \lor F_1 s_2 s_3 = s_3 \lor s_2 s_3 \lor s_1 s_2 s_3 \lor F_4 s_1 s_2 s_3 \\
    &= s_3 \lor s_2 s_3 \lor s_1 s_2 s_3 \lor s_1 s_2 s_3 s_4 \lor (s_1 s_2 s_3 s_4)^*; \\
F_4 &= s_4 \lor s_3 s_4 \lor F_2 s_3 s_4 = s_4 \lor s_3 s_4 \lor s_2 s_3 s_4 \lor F_1 s_2 s_3 s_4 \\
    &= s_4 \lor s_3 s_4 \lor s_2 s_3 s_4 \lor s_1 s_2 s_3 s_4 \lor (s_1 s_2 s_3 s_4)^*.
\end{aligned} \tag{3}
$$

Regular equation of business process in conditions of product production volume change has form:

$$
F = F_1 \lor F_2 \lor F_3 \lor F_4. \tag{4}
$$

Activity equation has form:

$$
\begin{aligned}
F = {} & s_1 \lor s_1 s_4 \lor s_1 s_3 s_4 \lor s_2 \lor s_1 s_2 \lor s_1 s_2 s_4 \lor s_3 \lor s_2 s_3 \lor s_1 s_2 s_3 \\
       & \lor s_4 \lor s_3 s_4 \lor s_2 s_3 s_4 \lor s_1 s_2 s_3 s_4 \lor (s_1 s_2 s_3 s_4)^*.
\end{aligned} \tag{5}
$$

Equation in algebra of relationship has form:

$$
\begin{aligned}
f_1 &= y_1 f_2; \\
f_2 &= y_2 f_3; \\
f_3 &= y_3 f_4; \\
f_4 &= y_4 f_1.
\end{aligned} \tag{6}
$$

Model activity in algebra of relationship has form:

$$
f_1 = (y_1 y_2 y_3 y_4)^*. \tag{7}
$$

So, the decision making process at the "growth" stage depends on all transitions functions, represented at the Fig. 3. These functions are performed sequentially.

The business processes sequence at the life cycle stage of "maturity" is represented at the Fig. 4. At this stage the improvement of equipment is implemented.

This sequence is represented by automate model, in which input signal is information about competitors (product cost, product quality etc.) and output signal is number of potential suppliers of equipment (Fig. 5).

Measures to ensure the competitiveness of products at the stage of "maturity", which is characterized by stable sales, are to reduce production costs and improve the quality of products to maintain their competitiveness, which is accompanied by the introduction of new or improvement of existing technological processes.

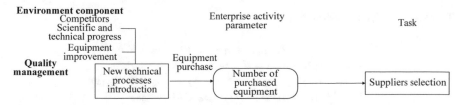

Fig. 4 Business processes at the stage of "maturity"

Fig. 5 Automate model of decision making process at the "maturity" stage

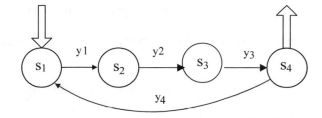

The input signal is information about competitors (for example, price and quality of products, etc.), and the output signal is the number of suppliers of equipment. The states (regular events) are:

- s_1—change of input information about competitors and scientific and technical progress achievements;
- s_2—necessary of equipment improvement;
- s_3—types and number of purchased equipment;
- s_4—number of equipment suppliers.

The transitions functions are:

- y_1—determination of purchased equipment number;
- y_2—determination of types and number of purchased equipment;
- y_3—determination of number of equipment suppliers;
- y_4—possibility of repetition in condition of environment change.

Process equations have the form:

$$F_1 = s_1 \lor F_4 s_1;$$
$$F_2 = s_2 \lor F_1 s_2;$$
$$F_3 = s_3 \lor F_2 s_3;$$
$$F_4 = s_4 \lor F_3 s_4. \tag{8}$$

Transform the process equations:

$$
\begin{aligned}
F_1 &= s_1 \vee s_1 s_4 \vee F_3 s_1 s_4 = s_1 \vee s_1 s_4 \vee s_1 s_3 s_4 \vee F_2 s_1 s_3 s_4 \\
&= s_1 \vee s_1 s_4 \vee s_1 s_3 s_4 \vee s_1 s_2 s_3 s_4 \vee (s_1 s_2 s_3 s_4)^*; \\
F_2 &= s_2 \vee s_1 s_2 \vee F_4 s_1 s_2 = s_2 \vee s_1 s_2 \vee s_1 s_2 s_4 \vee F_3 s_1 s_2 s_4 \\
&= s_2 \vee s_1 s_2 \vee s_1 s_2 s_4 \vee s_1 s_2 s_3 s_4 \vee (s_1 s_2 s_3 s_4)^*; \\
F_3 &= s_3 \vee s_2 s_3 \vee F_1 s_2 s_3 = s_3 \vee s_2 s_3 \vee s_1 s_2 s_3 \vee F_4 s_1 s_2 s_3 \\
&= s_3 \vee s_2 s_3 \vee s_1 s_2 s_3 \vee s_1 s_2 s_3 s_4 \vee (s_1 s_2 s_3 s_4)^*; \\
F_4 &= s_4 \vee s_3 s_4 \vee F_2 s_3 s_4 = s_4 \vee s_3 s_4 \vee s_2 s_3 s_4 \vee F_1 s_2 s_3 s_4 \\
&= s_4 \vee s_3 s_4 \vee s_2 s_3 s_4 \vee s_1 s_2 s_3 s_4 \vee (s_1 s_2 s_3 s_4)^*.
\end{aligned}
\tag{9}
$$

Regular equation of business process in conditions of product production volume change has form:

$$
F = F_1 \vee F_2 \vee F_3 \vee F_4.
\tag{10}
$$

Activity equation has form:

$$
\begin{aligned}
F = s_1 \vee s_1 s_4 \vee s_1 s_3 s_4 \vee s_2 \vee s_1 s_2 \vee s_1 s_2 s_4 \vee s_3 \vee s_2 s_3 \vee s_1 s_2 s_3 \\
\vee s_4 \vee s_3 s_4 \vee s_2 s_3 s_4 \vee s_1 s_2 s_3 s_4 \vee (s_1 s_2 s_3 s_4)^*.
\end{aligned}
\tag{11}
$$

Equation in algebra of relationship has form:

$$
\begin{aligned}
f_1 &= y_1 f_2; \\
f_2 &= y_2 f_3; \\
f_3 &= y_3 f_4; \\
f_4 &= y_4 f_1.
\end{aligned}
\tag{12}
$$

Model activity in algebra relationship has form:

$$
f_1 = (y_1 y_2 y_3 y_4)^*.
\tag{13}
$$

The sequence of business processes at the "decline" stage is obtained (Fig. 6).

The sequence of business processes at the stage of "decline" of products is due to a decrease in sales, which necessitates the introduction of product innovations in

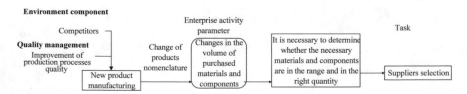

Fig. 6 Business processes at the stage of "decline"

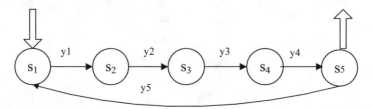

Fig. 7 Automate model of decision making process at the "decline" stage

order to maintain or search for new markets for products. At this stage the processes are possible:

- the manufactured product is being upgraded;
- the new types of product are manufactured.

To preserve the existing market share through marketing research, consumer preferences are assessed and the menstruation and quality indicators of new products are determined. Then planning of procurement of material and technical resources in the required volume is carried out and the task of selecting the suppliers of these resources is solved. The automate model of decision making process at the "decline" stage is represented in the Fig. 7.

The input signal is information about competitors (for example, price and quality of products, etc.), and the output signal is the necessity of new supplier's selection, which allows to ensure the production of change volumes of products, taking into account the quality management tasks.

The states (regular events) are:

- s_1—decision on the release of new types of products, made on the basis of the analysis of input information about competitors;
- s_2—organization of products quality control;
- s_3—change in the volume of production of new types of products;
- s_4—change of the required nomenclature and volume of purchased material resources;
- s_5—change in the number of material resources suppliers.

The transitions functions are:

- y_1—determination of the required material recourses number for new types of product;
- y_2—the quality control measures determination for new product;
- y_3—determination of the necessary nomenclature of purchased material resources;
- y_4—determination of the necessary number of material resources suppliers;
- y_5—possibility of repetition in condition of environment change.

Process equations have the form:

$$F_1 = s_1 \vee F_5 s_1;$$

$$F_2 = s_2 \vee F_1 s_2;$$
$$F_3 = s_3 \vee F_2 s_3;$$
$$F_4 = s_4 \vee F_3 s_4;$$
$$F_5 = s_5 \vee F_4 s_5. \tag{14}$$

The process of model action in algebra of languages can be represented:

$$F = F_1 \vee F_2 \vee F_3 \vee F_4 \vee F_5. \tag{15}$$

Process equations can be represented as follows:

$$
\begin{aligned}
F_1 &= s_1 \vee s_1 s_5 \vee F_4 s_1 s_5 = s_1 \vee s_1 s_5 \vee s_1 s_4 s_5 \vee F_3 s_1 s_4 s_5 \\
&= s_1 \vee s_1 s_5 \vee s_1 s_4 s_5 \vee s_1 s_3 s_4 s_5 \vee F_2 s_1 s_3 s_4 s_5 \\
&= s_1 \vee s_1 s_5 \vee s_1 s_4 s_5 \vee s_1 s_3 s_4 s_5 \vee s_1 s_2 s_3 s_4 s_5 \vee (s_1 s_2 s_3 s_4 s_5)^*; \\
F_2 &= s_2 \vee s_1 s_2 \vee F_5 s_1 s_2 = s_2 \vee s_1 s_2 \vee s_1 s_2 s_5 \vee F_4 s_1 s_2 s_5 \\
&= s_2 \vee s_1 s_2 \vee s_1 s_2 s_5 \vee s_1 s_2 s_4 s_5 \vee F_3 s_1 s_2 s_4 s_5 \\
&= s_2 \vee s_1 s_2 \vee s_1 s_2 s_5 \vee s_1 s_2 s_4 s_5 \vee s_1 s_2 s_3 s_4 s_5 \vee (s_1 s_2 s_3 s_4 s_5)^*; \\
F_3 &= s_3 \vee s_2 s_3 \vee F_5 s_2 s_3 = s_3 \vee s_2 s_3 \vee s_2 s_3 s_5 \vee F_4 s_2 s_3 s_5 \\
&= s_3 \vee s_2 s_3 \vee s_2 s_3 s_5 \vee s_2 s_3 s_4 s_5 \vee F_1 s_2 s_3 s_4 s_5 \\
&= s_3 \vee s_2 s_3 \vee s_2 s_3 s_5 \vee s_2 s_3 s_4 s_5 \vee s_1 s_2 s_3 s_4 s_5 \vee (s_1 s_2 s_3 s_4 s_5)^*; \\
F_4 &= s_4 \vee s_3 s_4 \vee F_2 s_3 s_4 = s_4 \vee s_3 s_4 \vee s_2 s_3 s_4 \vee F_1 s_2 s_3 s_4 \\
&= s_4 \vee s_3 s_4 \vee s_2 s_3 s_4 \vee s_1 s_2 s_3 s_4 \vee F_5 s_1 s_2 s_3 s_4 \\
&= s_4 \vee s_3 s_4 \vee s_2 s_3 s_4 \vee s_1 s_2 s_3 s_4 \vee s_1 s_2 s_3 s_4 s_5 \vee (s_1 s_2 s_3 s_4 s_5)^*; \\
F_5 &= s_5 \vee s_4 s_5 \vee F_3 s_4 s_5 = s_5 \vee s_4 s_5 \vee s_3 s_4 s_5 \vee F_2 s_3 s_4 s_5 \\
&= s_5 \vee s_4 s_5 \vee s_2 s_3 s_4 s_5 \vee F_1 s_2 s_3 s_4 s_5 \\
&= s_5 \vee s_4 s_5 \vee s_2 s_3 s_4 s_5 \vee s_1 s_2 s_3 s_4 s_5 \vee (s_1 s_2 s_3 s_4 s_5)^*.
\end{aligned}
\tag{16}
$$

The equation of functioning is:

$$
\begin{aligned}
F &= s_1 \vee s_1 s_4 \vee s_1 s_3 s_4 \vee s_2 \vee s_1 s_2 \vee s_1 s_2 s_4 \vee s_3 \vee s_2 s_3 \vee s_1 s_2 s_3 \\
&\quad \vee s_4 \vee s_3 s_4 \vee s_2 s_3 s_4 \vee s_1 s_2 s_3 s_4 s_5 \vee (s_1 s_2 s_3 s_4 s_5)^*.
\end{aligned}
\tag{17}
$$

The functioning of the model in the form of a regular expression of the algebra of relations can be represented:

$$f_1 = (y_1 y_2 y_3 y_4 y_5)^*. \tag{18}$$

Equations in algebra of languages determine sets of allowable values and describe changes of states of all elements of discrete system, which makes it possible to determine the sequence of performing tasks and converting relevant information in procurement quality management. The equations in the algebra of relations describe the possible states of the system, taking into account the conditions of transitions and control actions.

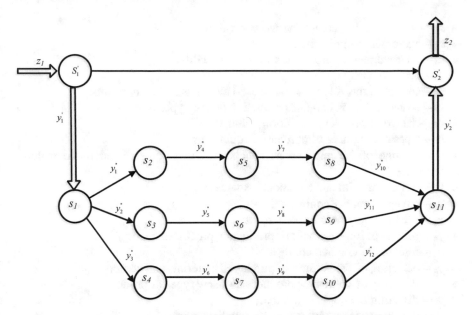

Fig. 8 Integrated model of quality management of logistics business processes based on discrete converter

The presence of management implies two interacting units—the control and controlled components, the interaction of which is described using the regular expression of relationship algebra, which allows taking into account the influence of the external environment on the result of solving the procurement management and quality management tasks.

The integrated model, based on automate models at different life cycle stages, is formed (Fig. 8). The model that describes the transition of the logistics system to different states under the influence of various control actions and changes in the parameters of the components of the environment is based on discrete converter model. It consists of two interacted components:

- the control component;
- the controlled component.

States and transitions of the control component are:

- S_1'—the need for changes in the quality management system of logistics business processes;
- S_2'—specific solutions to improve the quality of logistics business processes;
- y_1—solution of quality management logistics tasks.

States and transitions of the controlled component are:

- s_1—parameters value of quality management system of logistics business processes;

- s_2—need of quality control network creation;
- s_3—equipment improvement;
- s_4—improvement of production processes quality;
- s_5—demand change;
- s_6—change in the volume of purchased materials and components;
- s_7—change in the volume of production of new types of products;
- s_8—change of product production volume;
- s_9—types and number of purchased equipment;
- s_{10}—change of the required nomenclature and volume of purchased material resources;
- s_{11}—list of material and technical resources suppliers.
- y_1''—create of quality control network;
- y_2''—need to equipment improvement;
- y_3''—actions of improvement of production processes quality;
- y_4''—research of demand change;
- y_5''—detecting of the volume of purchased materials and components;
- y_6''—detecting of volume of production of new types of products;
- y_7''—detecting of suppliers number;
- y_8''—detecting of material resources suppliers number;
- y_9''—detecting of equipment suppliers number;
- y_{10}'', y_{11}'', y_{12}''—solution of the supplier selection problem.

Signals for interaction between the control and controlled components are:

- y_1'—management transfer;
- y_2'—transfer the information about suppliers;
- z_1—environment influence;
- z_2—contracting with suppliers.

The operation of the controlled component in the form of the algebra of algorithms regular expression has the following form:

$$f = y_1' y_4' y_7' y_{10}' \vee y_2' y_5' y_8' y_{11}' \vee y_3' y_6' y_9' y_{12}'. \tag{19}$$

Taking into account the influence of the external environment and the control component, the equation was obtained:

$$F = z_1 \left(y_1' f y_2' z_2 \vee y_1' z_2 \right). \tag{20}$$

Thus, according to the concept of Just-in-time in the management of procurement it is necessary to solve the following main tasks [15, 16]:

- redefinition of the need for materials and equipment in accordance with the volume of demand for manufactured products, which will determine the volume of production;

- reassessment of needs requires to define the need for MTP depending on the expected volume of production;
- task "make or buy" requires to define which types of material resources are more profitable to produce and which ones to purchase. When applying the principle of outsourcing at instrument-making enterprises, raw materials for the production of parts, some materials and components are purchased;
- determination of the type of procurement. It is necessary to determine from how many suppliers to purchase the necessary material resources;
- identification of all possible suppliers of a certain type of material resources that will meet the needs of the enterprise;
- selection of the most preferred suppliers, minimizing their quantity;
- minimization of stocks;
- solve tasks of quality management.

Thus, the integrated automaton model is proposed that describes the transition of a logistic system to different states under the influence of various control actions and changes in the parameters of the components of the external environment.

The options for implementing strategies at the stages of the product life cycle and the problems arising from the management of procurement of material and technical resources, taking into account the influence of the external environment, are considered. It was concluded that these tasks are associated with the initial stages of the life cycle, starting from the stage of marketing research. But most of the problems are solved at the stage of technological preparation of production. Therefore, it became necessary to determine the main parameters and the place of logistic tasks for different stages of the product life cycle: "growth, maturity, recession". Thus, at each stage, the management goal, quality management, the necessary changes in the production and logistics system and the corresponding control actions are determined.

For a formal presentation of the procurement management tasks, it is proposed to use an automaton model apparatus, which allows to determine the sequence of actions when making decisions. For each of the three selected stages, the features of procurement management decision-making were considered, which made it possible to form the corresponding state models.

In order to form an invariant information technology of decision-making, an integrated automaton model was developed, which describes the transition of a logistic system to different states under the influence of various control actions and changes in the parameters of the components of the environment taking into account quality management tasks.

1.2 Development of Environment Influence Evaluation Method

It's proposed to evaluate the influence of environmental factors (z_1, z_2) on the business processes of the manufacturing enterprise [16], using the full factor experiment (FFE) [17] (Fig. 9).

The manufacture enterprise system components are:

- quality management;
- sale;
- manufacture;
- supply.

In carrying out measures to ensure the quality of products required:

- identify the set of factors that significantly affect the resulting values of the quality of the intermediate product at the output of the i-th element of the logistics chain;
- conduct a predictive modeling of quality management activities in accordance with the full-factor plan of the experiment;
- obtain the analytical dependence (regression) from which it's possible to determine the significance of factors that affect the quality of the product for the i-th element

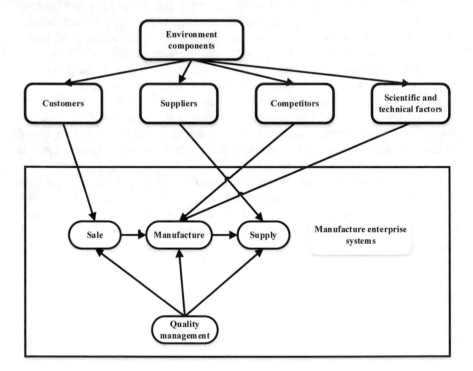

Fig. 9 Environment influence to manufacture enterprise systems

of the logistics chain and the direction of change of these factors (decrease or increase). Factors that have the least impact can be discarded;

- conduct an optimization to determine such a combination of factors that improve quality and implemented value K_i^{i-1}.

Sale, manufacture and supply are components of logistic system of production manufacture. According to a simple plan of a full factor experiment, experts in quality distinguish three factors $x_{1_i}, x_{2_i}, x_{3_i}$ that affect the quality of the product at the output of the i-th element of the logistics chain. For FFE, each factor must be represented in the range in the form of minimum and maximum values.

The incomplete-rational regressive dependence of the species is constructed [18, 19]:

$$
\begin{aligned}
K = k_0 &+ k_1 \cdot y_1 + k_2 \cdot y_2 + \cdots + k_n \cdot y_n \\
&+ k_{12} \cdot y_1 \cdot y_2 + \cdots + k_{123} \cdot y_1 \cdot y_2 \cdot y_3 + \cdots \\
&+ k_{123\ldots n} \cdot y_1 \cdot y_2 \cdot y_3 \ldots y_n,
\end{aligned}
\tag{21}
$$

where the value k_i indicates the degree and direction of the external factors influence on the quality performance of business processes K.

The results of the calculation and analysis of the factors influence using the FFE was obtained. These results are used to activities plan that are carried out to ensure the quality of produced products. For a large number of possible measures, the choice is made using the method of integer linear programming [19].

To ensure the quality of products, the enterprise must plan and forecast [20] quality measures that address both the main and auxiliary processes of production (automation of the technological process, equipment modernization, strengthening of entrance control, improvement of industrial discipline, improvement of personnel qualification, etc.) [20–22]. To carry out events it is necessary to allocate funds and to control the content and terms of execution. Use such denotation:

- the j-th possible event in the form M_j;
- terms of execution—T_j;
- required resources (including limited)—C_j;
- the risk of successful execution—R_j.

Use the boolean variable $y_{ij} \in \{0; 1\}$, which means that if measures in the i-th plan of measures M_j used, then $y_{ij} = 1$. Otherwise $y_{ij} = 0$, where $i = \overline{1, m}$, m—number of possible measures, $m \leq n$. As the value of the target function, we will use the values K_i of quality that are asked by experts when sets of measures are considered.

It's need to maximize:

$$
\max_i K_i, \quad i = \overline{1, N'}, N' = 2^m
\tag{22}
$$

when performing the following restrictions related to the limited resources of the enterprise:

$$C_i = \sum_{j=1}^{m} c_{ij} y_{ij}, \quad C_i \le C';$$

$$T_i = \sum_{j=1}^{m} t_{ij} y_{ij}, \quad T_i \le T'; \tag{23}$$

$$R_i = \sum_{j=1}^{m} r_{ij} y_{ij}, \quad R_i \le R';$$

where C', T', R'—limits are related to permissible values of cost, time and risk when choosing directions for improving quality;

c_{ij}—expenses for the implementation of the j-th event with the choice of the i-th direction of the quality improve;

t_{ij}—time spent on the j-th event when choosing i-th direction of quality improvement;

r_{ij}—the risk connected with the implementation of the j-th event when choosing i-th direction of quality improvement.

When assigning values c_{ij} it's desirable to take into account the significance of factors (measures), in the form of k_j coefficients constructed with FFE. For convenience, the coefficients k_j can be normalized:

$$\hat{k}_j = \frac{k_j}{\sum_{j=1}^{m} k_j}, \quad \sum_{j=1}^{m} \hat{k}_j = 1. \tag{24}$$

Taking into account the limited resources of the enterprise, the following distribution of resources were obtained:

$$C_i = \sum_{j=1}^{m} \hat{k}_j \cdot C' \cdot y_{ij}, \quad C_i \le C'. \tag{25}$$

In the case of parallel execution of the measures, in order to assess the time of execution of the i-th direction:

$$T_i = \max_j t_{ij} y_{ij}, \quad T_i \le T'. \tag{26}$$

It is possible to use an approach related to minimization of resources without compromising the necessary quality of products. In this case it isn't necessary to find:

$$\min_i C_i, \quad C_i = \sum_{j=1}^{m} c_{ij} y_{ij}, \tag{27}$$

with the following limits:

$$K_i \geq K';$$

$$T_i = \sum_{j=1}^{m} t_{ij} y_{ij}, \quad T_i \leq T'; \qquad (28)$$

$$R_i = \sum_{j=1}^{m} r_{ij} y_{ij}, \quad R_i \leq R';$$

where K'—the required level of quality that ensures the competitiveness of the manufactured products.

Reducing the number of factors (measures) affects the quality of the product which allows reducing the dimension of the solvable problem, which influences, in the future, to minimize the time, money, and risk when choosing strategies for improving quality [18–24]. Quality assessments, obtained with the help of experts, are new to the preparation of a quality plan. Upon expiration of the plan, it is necessary to evaluate the differences between actual and planned quality values. The following situations are available:

- planned (K_p) and actual (K_a) values of quality are close $K_p \approx K_a$ and the achieved results satisfy the management of the enterprise;
- planned and actual values of quality are very different $K_p \geq K_a$.

In the first case, if the goal is achieved, then new means for improving the quality will not be allocated or will be allocated in a limited number to those measures that led to success. In the second case, it is necessary to gather a team of experts (it is desirable to update it in order to conduct a reliable assessment) and to conduct a quality assessment, reviewing, with the help of experts, a set of measures. Then solve two successive tasks and form the plans for implementing measures to improve quality.

So, the method of logistic production management has been improved, which implements the principle of phased quality assurance and uses the results of logistics chain modeling, which allows optimizing the limited resources of the enterprise to improve production.

The value of product quality corresponds to the set of measures on all elements of the logistics chain, from the first to the last. it allows to use the limited possibilities of the enterprise and the allocated funds for quality assurance, as well as with the actual systemic dependence of the quality value of the intermediate product for the i-th element of the logistic chain with the future value of the quality of the product, which is formed by carrying out activities at a subsequent $i + 1$ element of the logistic chain of production.

The following strategy of consistent improvement of quality is proposed.

1. Analyze the possibility of improving the quality of the first element of the logistics chain. Determine the value K_1 and ΔK_1. Then a number of measures M_1 are planned for implementation ΔK_1.

2. As a result of the measures taken to improve the quality, the actual values K_1 is obtained in the form K_{1f} and ΔK_{1f} for the first element of the logistics chain.
3. Taking into account the received actual values K_{1f} and ΔK_{1f} and the quality experts review the requirements for production quality for the following, after the first, elements of the logistics chain:

$$K_2^1, K_3^1, \ldots, K_n^1;$$

$$\Delta K_2^1, \Delta K_3^1, \ldots, \Delta K_n^1. \tag{29}$$

1. In the second element of the logistics chain the set of measures M_2 for implementation ΔK_2^1 are obtained.
2. As a result of the measures taken to improve the quality, we obtain the actual values K_2^1 in the form K_{2f}^1 and ΔK_{2f}^1 for the second element of the logistics chain.
3. Taking into account the obtained values K_{2f}^1 and ΔK_{2f}^1 quality experts review the quality requirements following the second element of logistics chain:

$$K_3^2, K_4^2, \ldots, K_n^2;$$

$$\Delta K_3^2, \Delta K_4^2, \ldots, \Delta K_n^2. \tag{30}$$

For the third and subsequent elements of logistics chain are repeated. For the i-th element of the logistics chain is obtained:

$$K_i^{i-1}, K_{i+1}^{i-1}, \ldots, K_n^{i-1};$$

$$\Delta K_i^{i-1}, \Delta K_{i+1}^{i-1}, \ldots, \Delta K_n^{i-1}. \tag{31}$$

The final actual value of product quality is determined by the last, initial element of the logistics chain and corresponds to the set of measures carried out on all elements of the logistics chain, from the first to the last.

2   Conclusion

Ensuring the competitiveness of domestic products in foreign markets, primarily associated with improving the quality of products in the aerospace industry. The limited resources of the enterprise, as well as the complexity of obtaining investment, lead the management to solve the quality problem in the light of its limited capabilities.

The equations of processes in the algebra of languages determine the set of allowable values and describe the changes in the states of all elements of a discrete system, which makes it possible to determine the sequence of performing tasks and converting relevant information in procurement and quality management. The equations in the algebra of relations describe the possible states of the system, taking into account the conditions of transitions and control actions.

The business processes at the stages of the products life cycle and the tasks of managing the procurement of materials and equipment, taking into account the quality management tasks and environmental impact are considered. The goal of management, the quality management task, the necessary changes in the production and logistics system and the corresponding control actions are defined at each stage.

For the formal presentation of procurement management tasks, it was proposed to use an automaton model apparatus, which allows determining the sequence of actions for decisions making support. For each of the four selected stages, the features of procurement management decision-making were considered, which made it possible to form the corresponding state models.

The optimization model based on the method of full-factor experiment is proposed. This model estimates the influence of environmental factors in the conditions of the limited enterprise resources. The method of planning measures to ensure the quality of products is offered. A set of possible measures is represented in the form of a combination of zeros and units. The complete set of measures corresponds to a full-featured experiment, which using the experts' knowledge can identify significant factors (measures) that affect the quality. For a large number of possible measures an optimization model has been created for choosing a plan of activities in the conditions of limited resources of the enterprise. The optimization model for substantiation of the multilevel quality control system is developed.

The following scientific results were obtained:

- the method of logistic business processes management based on discrete converter is improved;
- the method of logistic management is improved, which implements the principle of stage-by-stage quality assurance and uses the results of modeling of the logistic chain, which allows to optimize the limited resources of the enterprise for improvement of production;
- the method of production development planning is improved by automated expert evaluation and a full-factor experiment to substantiate and select measures to ensure the quality of high-tech products of the enterprise.

References

1. Kryvinska N (2012) Building consistent formal specification for the service enterprise agility foundation. Soc Serv Sci J Serv Sci Res 4(2):235–269 (Springer)
2. Kaczor S, Kryvinska N (2013) It is all about services—fundamentals, drivers, and business models. Soc Serv Sci J Serv Sci Res 5(2):125–154 (Springer)

3. Gregus M, Kryvinska N (2015) Service orientation of enterprises—aspects, dimensions, technologies. Comenius University in Bratislava
4. Kryvinska N, Gregus M (2014) SOA and its business value in requirements, features, practices and methodologies. Comenius University in Bratislava
5. Molnar E, Molnar R, Kryvinska N, Gregus M (2014) Web intelligence in practice. Soc Serv Sci J Serv Sci Res 6(1):149–172 (Springer)
6. Kirichenko L, Radivilova T, Zinkevich I (2017) Forecasting weakly correlated time series in tasks of electronic commerce. In: Proceedings of 2017 12th international scientific and technical conference on computer sciences and information technologies (CSIT), Lviv, pp 309–312. https://doi.org/10.1109/stc-csit.2017.8098793
7. Kirichenko L, Radivilova T, Zinkevich I (2018) Comparative analysis of conversion series forecasting in e-commerce tasks. In: Shakhovska N, Stepashko V (eds) Advances in intelligent systems and computing II. CSIT 2017. Advances in intelligent systems and computing, vol 689. Springer, Cham, pp 230–242. https://doi.org/10.1007/978-3-319-70581-1_16
8. Dobrynin I, Radivilova T, Maltseva N, Ageyev D (2018) Use of approaches to the methodology of factor analysis of information risks for the quantitative assessment of information risks based on the formation of cause-and-effect links. In: Proceedings of the 2018 international scientific-practical conference problems of infocommunications. Science and technology (PIC S&T). Kharkiv, Ukraine, pp 229–232. https://doi.org/10.1109/infocommst.2018.8632022
9. Kirichenko L, Bulakh, V, Radivilova, T (2017) Fractal time series analysis of social network activities. In: Proceedings of the 2017 4th international scientific-practical conference problems of infocommunications. Science and technology (PIC S&T). Kharkov, Ukraine, pp 456–459. https://doi.org/10.1109/infocommst.2017.8246438
10. Kirichenko L, Radivilova T, Tkachenko A (2019) Comparative analysis of noisy time series clustering. In: Proceedings of the 3rd international conference on computational linguistics and intelligent systems, vol I (COLINS-2019). Kharkiv, Ukraine, 18–19 April 2019, pp 184–196
11. Ageyev D (2010) NGN network planning according to criterion of provider's maximum profit. In: Proceeding of the IEEE international conference on modern problems of radio engineering, telecommunications and computer science (TCSET-2010), pp 256–256
12. Ageyev D, Al-Ansari A, Qasim N (2015) Multi-period LTE RAN and services planning for operator profit maximization. In: Proceeding of the IEEE the experience of designing and application of CAD systems in microelectronics, pp 25–27. https://doi.org/10.1109/cadsm. 2015.7230786
13. Ageyev D, Wehbe F (2013) Parametric synthesis of enterprise infocommunication systems using a multi-layer graph model. In: Proceeding of the IEEE 23rd international crimean conference microwave and telecommunication technology, pp 507–508
14. Al-Dulaimi A, Al-Dulaimi M, Ageyev D (2016) Realization of resource blocks allocation in LTE downlink in the form of nonlinear optimization. In: Proceeding of the IEEE 13th international conference on modern problems of radio engineering, telecommunications and computer science (TCSET), pp 646–648. https://doi.org/10.1109/tcset.2016.7452140
15. Dekker R, Fleischmann M, Inderfurth K, Van Wassenhove LN (2013) Reverse logistics. Quant Model Closed-Loop Supply Chain. https://doi.org/10.1007/978-3-540-24803-3
16. Jung KS, Dawande M, Geismar HN, Guide VDR Jr, Sriskandarajah C (2016) Supply planning models for a remanufacturer under just-in-time manufacturing environment with reverse logistics
17. Quality management principles. Homepage. http://www.iso.org/iso/qmp_2012.pdf Last accessed 05 May 2019
18. Juran JM, (2004) Architect of quality. The autobiography of Dr. Joseph M. Juran. McGraw-Hill, New York
19. George U (2008) Use of statistical techniques in quality management systems. In: The 8th international conference "reliability and statistics in transportation and communication (RelStat'08)", pp 329–334. Riga, Latvia
20. Croft NH, ISO 9001:2015 and beyond—preparing for the next 25 years of quality management standards. Homepage. http://www.iso.org/iso/home/news_index/news_archive/news. htm?refid = Ref1633. Last accessed 28 Oct 2018

21. Outlook on the logistics and supply chain industry 2013. Homepage. http://www3.weforum.org/docs/WEF_GAC_LogisticsSupplyChainSystems_Outlook_2013.pdf. Last accessed 10 Jan 2019
22. Barcik RA (2013) The significance of customer service in logistics. J Logist Transp 1(17):5–10
23. Mathauer M, Hofmann E (2019) Technology adoption by logistics service providers. Int J Phys Distrib Logist Manag 49(4):416–434. https://doi.org/10.1108/IJPDLM-02-2019-0064
24. Talay I, Özdemir-Akyıldırım Ö (2019) Optimal procurement and production planning for multi-product multi-stage production under yield uncertainty. Eur J Oper Res 275:536–551

Machine Learning Methods Applications for Estimating Unevenness Level of Regional Development

Liubov Chagovets⊙, **Vita Chahovets**⊙ and **Natalia Chernova**⊙

Abstract The article deals with the issues of regional socio-economic development. Particular attention is paid to the estimation and analysis of the level of unevenness on different levels of hierarchy. Some international ratings of countries are considered and the place of Ukraine is determined. It is proved that the relatively low position of the country is determined by lots of different external and internal factors, but one of the most significant ones is the regional unevenness of development. From the other hand existing disparities in the development of particular regions can play the role of incentives and to contribute to the improvement of the situation in the country. The crucial condition of such improvement is high-quality and grounded control signals. Implementation of inefficient control signals may deepen the gaps between the levels of regional development. The main aim of the research is to construct the complex of models for estimation and forecasting the unevenness level of social and economic development of regions. The complex is based on methods of modeling of multidimensional objects. The proposed models are united in three consecutive blocks. The aim of the fist block is to conduct a priory analysis of the initial set of socio-economic indicators and to form the representative indicator system. The second block applies methods of multidimensional analysis. The third block represents models of forecasting of the structural components of the unevenness level. The proposed models should be implemented in the decision-making process and to became the foundation for efficient regional socio-economic policy.

L. Chagovets (✉) · N. Chernova
Simon Kuznets Kharkiv National University of Economics, 9a Nauki av., Kharkiv 61166, Ukraine
e-mail: chahovets.liubov@hneu.net

N. Chernova
e-mail: natacherchum@gmail.com

V. Chahovets
Kharkiv State University of Food Technology and Trade, 333 Klochkivska Str., Kharkiv 61051, Ukraine
e-mail: chagovec.v@ukr.net

© Springer Nature Switzerland AG 2020
D. Ageyev et al. (eds.), *Data-Centric Business and Applications*,
Lecture Notes on Data Engineering and Communications Technologies 42,
https://doi.org/10.1007/978-3-030-35649-1_6

Keywords Region · System · Unevenness · Socio-economic development ·
Indicator · Model · Estimation · Cluster · Representative · Adaptive forecasting ·
Management

1 Introduction

Nowadays level of social and economic development of separate regions in Ukraine
is determined by a number of external and internal factors, that have destabilizing
and stochastic nature. Sustainable and balanced growth of all regions is the key factor in maintaining strategic priorities of the country and ensuring national safety.
High degree of unevenness causes deep disproportions between regions. The disproportions are observed in the majority of spheres (economic, financial, social,
demographic, political, ecological and so on). Among the most crucial ones are differences in the industrial structure and unequal rates of development of the main
brunches. These factors greatly increase the risks and the likelihood of a loss of the
viability of the country's economic system, rise competitive pressure on international
trading markets, causes general economic downturn and inefficiency of the world
regional complexes functioning.

From the other hand existing disparities in the development of separate regions can
play the role of incentives to improve the current state of the recipient regions. The
crucial condition of such improvement is high-quality and grounded control signals.
Implementation of inefficient control signals on different levels of hierarchy may
deepen the gaps between the levels of socio-economic development of the regions.

Thus complex mathematical modeling of uneven social and economic development of the regions is the relevant actual research task especially from the point of
view of the development of regional policy, that ensures balanced development of
regions and the country as a whole. Moreover, the large variety of subjects, complexity and multiplicity of tasks in regional management cause absence of generally
accepted method of estimating the unevenness of socio-economic development of
regions.

The main aim of the research is to construct the complex of models for estimation and forecasting the unevenness level of social and economic development of
regions. The complex is based on methods of economic-mathematical modeling of
multidimensional objects. The application of the models can improve the quality of
decisions in regional management.

2 Literature Review

The analysis of scientific literature [5–15, 18–24, 26, 29, 30, 32–34, 39–42, 46, 47,
50–53, 55–61], devoted to the problems of uneven development of regions, have
shown some key definitions of the term "unevenness". These definitions are quite

similar in their content. They differ from each other by the specifics, peculiarities, a certain direction and sphere of functioning within which the category is considered. In many studies, "uneven development" is identified with "differences". Therefore, the "uneven development of regions" is determined as the existence of differences in a certain set of parameters that reflect the areas of functioning and development of regions.

Today the problem of regional inequality is the subject of active discussion and is widely considered in the scientific literature. Along with the notion of "unevenness", the terms "disproportionality", "asymmetry", "polarization", "differentiation" and "divergence" are often mentioned [9–11, 22, 59]. All of them carry a similar payload. We agree with the idea [8], where "differentiation", "polarization", "disproportion", "disharmony" and "enclave" are referred to as "forms of manifestation or degrees of inequality". There is also an opinion [11] that there is a connection between the concepts of "unevenness", "differentiation", "asymmetry" and "polarization". This connection is manifested in the following. The last three concepts characterize "unevenness"; the change in unevenness takes place in three stages: "differentiation— asymmetry—polarization"; at each of these stages an intensification of differences takes place; the fact of intensification allows to assert that each stage corresponds to a certain degree of uneven development; the deepening of the differences is caused by the impact of a set of certain factors.

According to researchers [18, 23], "differentiation" is the process of formation and development of inconsistency between regions, which is determined by differences in conditions, factors and results of regional development. The authors emphasize that "differentiation" is one of the laws of economic development, which leads to imbalance and disproportion in the development of regions. In some researches so-called deviations (shifts, gaps) in the parameters that describe the state of the regional economic system are associated with the term "asymmetry".

"Asymmetry" is a deviation in the conditions and results of the development of the social and economic spheres of regions. The same view is held by the authors [10, 16, 18, 23, 24, 26, 56]. They link "asymmetry" with "gaps" (deviations) in the performance of regions.

This testifies that due to the actions of certain factors the unevenness of regional development passes from the stage of "differentiation" to the stage of "asymmetry". The differences of the "differentiation stage" violate the integrity of the space. At the asymmetry stage, destabilization of reproductive processes becomes more significant. Finally, the polarization occurs.

Thus, the comparison and analysis of the concepts of "differentiation", "asymmetry" and "polarization" shows that each of them is associated with differences that characterize the uneven social and economic regional development.

In order to prevent the negative consequences of this phenomenon, an active participation of the government is necessary in the territorial development regulation.

Table 1 Legally defined algorithms of assessing interregional differentiation in Ukraine

Contents of method	Formula of calculation	Explanation to the formula
The algorithm assesses the differences between the most prosperous and the most problematic regions	$$C_i = \frac{\max_i(x_i)}{\min_i(x_i)}$$	C_i—the value of difference; $\max_i(x_i), \min_i(x_i)$ —maximum (minimum) value of the ith indicator
The algorithm assesses the range of deviation of the regional indicators from their average value	$$\bar{x} = \frac{1}{n}\sum_{j=1}^{n} x_j;$$ $$C_B = \sqrt{\frac{1}{n}\sum_{j=1}^{n}(x_j - \bar{x})^2};$$ $$K_B = \frac{C_B}{\bar{x}}$$	$\bar{x}$—average value; x_j—value of the jth indicator; n—number of regions; C_B—standard deviation; K_B—coefficient of variation

Today in Ukraine the only legally established methodology for assessing regional development unevenness is the following document [52]. According to this document two assessment algorithms are applied on the annual basis. The first algorithm assesses the differences between the most prosperous and the most problematic regions. The second algorithm assesses the range of deviation of the regional indicators from their average value (see Table 1).

Let's discuss both algorithms mentioned above.

As for the first algorithm it should be noted that the value of difference between the most prosperous and the most problematic regions gives only a generalized description of the differentiation level.

The first algorithm does not show the location of other regions relative to the corresponding indicators in the most prosperous and in the most problematic ones.

As for the second algorithm it should be noted that according to [60] the level of interregional differentiation of socio-economic development depends on the value of the coefficient of variation (the greater the value of this coefficient, the greater the degree of differentiation). It is necessary to note that the coefficient of variation is a quantitative characteristic of statistical data and is interpreted as a relative indicator of the difference in the values of a certain indicator that describes different objects from a certain set [11]. The usage of the variation complicates the objectivity of determining the level of regional unevenness because it is determined only by partial values of the average deviation of the whole system [28]. In addition to the above arguments the coefficient of variation cannot provide the meaningful interpretation of its values in dynamics in terms of regional unevenness. That is, assume that the value of the coefficient of variation in the current year is lower than in the previous year. This result may be interpreted as positive change. But it is impossible to make a choice between two alternative reasons of such change (accelerated pace of development of problem regions or decline in economic development in leading regions).

Taking into account the given contradictions of the algorithms, the methodological approach for assessing interregional disproportionality of social and economic development in the country needs to be improved.

Similar issues have been covered in various aspects in a large number of scientific works [5–61].

The core peculiarities of these approaches are presented in the Table 2.

Table 2 Methodological approaches review

Author, system of indicators	Calculation formula	Advantages of the technique	Disadvantages of the technique
B. Lavrovskij—main socio-economic indicators of development	$K_{var}^x =$ $\frac{1}{\lambda^x} \times \times \sqrt{\frac{1}{N} \sum_{i=1}^{N} (\lambda_i^x - \lambda^x)^2}$ K_{var}^x—coefficient of variation in the t-th year; λ^x—mean per capita value of indicator; λ_i^x—per capita value of indicator for i-th region; N—number of regions	Allows to conduct the analysis in dynamics; takes into account the indicators of each region	It is the basic coefficient of variation, adapted to economic indicators in a regional context
I. Storonyanska—gross regional product (GRP), population of the region	Tail's index: $IT = \sum_{i=1}^{i} \frac{Y_i}{Y} \cdot \ln \frac{Y_i/P_i}{Y/P},$ Y_i—GRP, Y—sum of GRP, P_i—population of the region, P—population of the country	Covers the total number of regions convenient in conducting calculations	Indicators of innovation, labor market, foreign trade activity are not taken into account, the obtained result is difficult to interpret
Aloïs Kotcher'er—main socio-economic indicators of regional development	$U_{ij} = \frac{x_{ij} - \max_i(x_i)}{S_{xi}};$ $U_{ij} = \frac{\min_i(x_i) - x_{ij}}{S_{xi}}$ x_{ij}—value of the ith indicator for the jth region; $\max(x_i)$—maximum value of the ith indicator; $\min(x_i)$—minimum value of the ith indicator; S_{xi}—standard deviation of the ith indicator	Peculiarities of such indicator types as stimulators and stimulators are taken into account; convenient in conducting calculations	Impossibility of application when the minimum value of indicator is equal zero
T. S. Klebanova—main socio-economic indicators of regional development	$k_g = \frac{I_{ij}}{\max(I_{ij})},$ k_g—coefficient, I_{ij}—integral development indicator for the region at the jth period	The possibility of assessing the asymmetry of development unevenness in dynamics for each region	It is impossible to determine the total unevenness level for the whole country

(continued)

Table 2 (continued)

Author, system of indicators	Calculation formula	Advantages of the technique	Disadvantages of the technique
V. I. Shevcova—the overall level of economic development of regions, the labor market, incomes and investment activity	$R_j = \frac{P_{jmax}}{P_{jmin}}$, R_j—imbalance level for the jth indicator, P_{jmax} and P_{jmin}—maximum and minimum values for the jth indicator	Simple practical application; economic and social indicators are taken into account separately	The system of indicators does not take into account the level of innovations, demographic characteristics, the total number of regions, the intermediate values of the indicators

It is important to specially note the unevenness index proposed by Klebanova [40]. It represents the total unevenness situation for all regions more precisely. This coefficient is calculated for each region as a fraction of its integral taxonomic index from the same maximum indicator in a set:

$$k_{ij} = \frac{I_{ij}}{\max\limits_{i=1,\dots,n}(I_{ij})}; j = \overline{1, m},\tag{1}$$

I_{ij} —unevenness index for the i region at the j period;
m —number of periods;
n —number of regions

The conducted analysis of the peculiarities of these methodical approaches allows us to understand and generalize their main advantages.

We see that authors use different methods and algorithms for assessing socio-economic unevenness between regions. Some of them propose to calculate the total level of unevenness for the panel data set. Other ones obtain the individual levels of unevenness for each region and then compare those levels.

As a result of the conducted analysis we can generalize the core advantages of these methods and algorithms. Most methods suggest the synthetic index of socio-economic development, which ensures the integrity of the results. The absence of expert assessments increases the objectivity of the obtained values. The possibility of conducting an estimation in dynamics allows to make a comparative analysis at different time intervals.

The main disadvantages are the following. Clear unified system of indicators for assessing the level of disproportionality is not determined. This fact leads to the difficulties in comparing the obtained results. The peculiarities of the calculation formulas complicate the adaptation of these methods to the existing interregional imbalances in Ukraine. The relationships between the obtained results and changes in the rates of development of regional economies are not established. The period during which the evaluation should be carried out is not defined. The direction of regional processes (convergence or divergence) in the country is not defined clearly. All these facts reduce the effectiveness of decision-making processes in regional management.

So, one of the research tasks is to improve the methodological approach to the assessment of interregional disproportionality. The pros and cons of the algorithms mentioned earlier should be taken into account whenever possible. So, the following points are highlighted for further study:

Firstly, the assessment of interregional disproportionality should be conducted using the integral index of economic development. It should be calculated for each region. The calculation algorithm shouldn't be based on expert methods.

Secondly, the initial system of indicators should contain such ones, that are taken from the official statistics. That is, the system should be informationally available.

Thirdly, the algorithm of the analytical part of the research should be clearly formulated. Time interval during which the unevenness dynamics is investigated should be determined. This will facilitate conducting such studies by other specialists in the future.

Fourthly, it is recommended to conduct comparative research of the unevenness level and the rates of growth of the main indicators of economic development in the regions. Such approach ensures completeness and complexity of the research.

Fifthly, the main directions of regional processes in the country should be determined. They will become the basis for the formation and adjustment of regional policy measures in the sphere of disproportionality management. As a result, the total level of disproportionality must be decreased.

3 Problem Formulation

In order to develop the complex of models for estimating, forecasting and managing the regional socio-economic unevenness the methods of system analysis should be applied.

The system analysis is based on certain principles [1–4, 13, 45, 51, 53]:

the possibility to split the system into separate parts - subsystems;
the requirement to consider the set of individual elements as a single system, rather than simply combining them;
the system has special properties that may not be represented in separate elements;
the existence of fundamental value function. This function represents the dependencies between the efficiency of the system and its structure;
the influence of the external environment of the system on its functioning. In other words, the current system is researched as a part of another system.

In practice, the system approach is system coverage, system representation, system orientation of research. It requires problem research in various aspects and from different positions.

In the most general form, any system is a converter of input and output impacts. This can be given as follows:

$$R : X \cdot C \to Y, \tag{2}$$

R —global reaction;
X —set of inputs;
C —set of states of the system;
Y —set of outputs.

"Concept" is a central system-creating element of a scientific research. Conceptual provisions and other constructions that specify it—structural elements of the concept.

Thus, the conceptual scheme of the research is proposed below (see Fig. 1). It presents the research logic and shows the dependencies between models.

Let's consider the conceptual scheme in more detail. According to the Fig. 1, the main methods that are applied within the research are: methods of spatial statistical analysis, multidimensional analysis, methods of adaptive forecasting.

The conceptual scheme consists of three consecutive blocks of models. Each block is described in detail below.

The main aim of the fist block is to conduct a priory analysis of the initial set of socio-economic indicators. The initial indicators represent not internal situation in the region but also allow make some kind of comparative analysis. Such research compares the situation in different regions within the country and in the regions from different countries based on international ratings. As a result, strengths and weaknesses of the regional socio-economic system may be distinguished. Finally, set of indicators that describe the regional unevenness has to be formed.

The second block applies methods of multidimensional analysis.

The first model within this block allows to classify regions on separate classes based on their level of socio-economic development. It applies methods of cluster analysis and solves the following tasks: comparison of research results according to different rules of aggregation; estimation of the quality of clusterization by dendrograms; determine the final number of clusters; application of iterative cluster methods; determine cluster characteristics; analysis of classification results for different number of isolated clusters; analysis of the main statistics and estimation of the significance of the variables by the constructed models.

The second model estimates loadings of representative indicators of unevenness. It applies methods of partial reduction and factor analysis. The main aim of the model is to reduce the dimension of the initial indicator set and to produce a new set of factors.

The model determines minimum required number of principal components based on cumulative dispersion analysis, analyze the values of factor loadings, detect latent factors that influence the level of unevenness.

The third model applies classification trees technique for recognition of the unevenness class of regional socio-economic development. It takes into consideration methods of discriminant one-dimensional branching, discriminant multidimensional

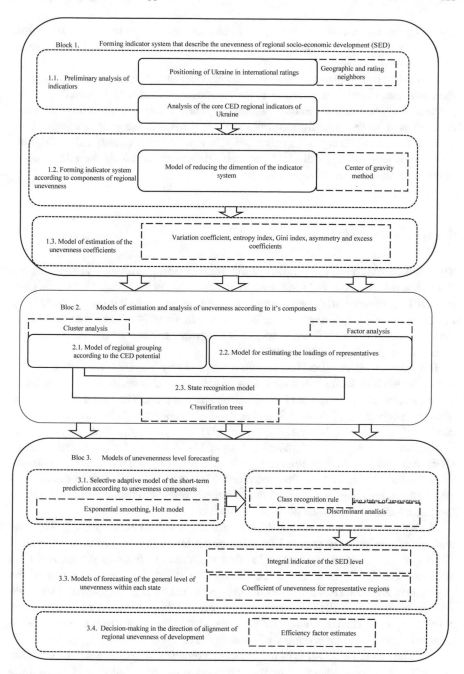

Fig. 1 Conceptual scheme

branching by linear combinations of variables and branching by CART algorithm. The final logical classification rules are formed as a result of comparative analysis of constructed models.

The third block represents models of forecasting of the structural components of the unevenness level. The selective adaptive models of short-term forecasting are the models of exponential smoothing and Holt models. Class recognition of predicted data is based on the classification rules that are formed in the previous steps. Then a set of decision-making rules is formed. They aimed to decrease and smooth the unevenness level of regional socio-economic development. Lastly the effectiveness of the proposed models is assessed.

4 Methods, Findings

The implementation of the proposed models involves a preliminary analysis of initial indicators and essential factors of socio-economic unevenness. One of the most crucial is the factor of the environment. Other regions of Ukraine, some regions of foreign countries, foreign countries as a whole should be taken into consideration during the research. That is why it is necessary to conduct the analysis of the international positions of the country.

For instance, the appropriate research was held by VoxUkraine company. The list of research tasks included estimates of some global indexes for a given set of countries and positioning Ukraine within obtained ratings. The initial set of indicators was analyzed and divided into five groups: "economics", "liberties", "human capital", "corruption", "state and security" [57]. The first group consists of indexes of economic development of countries. It should be noted that for Ukraine the lack of significant progress in economic ratings is not a surprise. The Index of Competitiveness, the Index of Economic Freedoms or the Index of Prosperity are systemic ones and cover many areas that are not always directly related to the economics. Ukraine has the middle or the lowest positions in such ratings.

A steady increase in incomes is the main reason for the economic growth of regions and the whole state. Household incomes characterize the dynamics of individual blocks of the economic system and reflect the state of affairs in the area of final consumption. It allows to assess the situation of improving the well-being of the population. As for Ukraine, the following regions had the highest level of disposable income per person in 2017: Dnipropetrovsk, Zaporozhye, Kiev, Odesa, Kharkiv and Poltava regions. The core factor affecting the size of the population's income is the sectoral structure of the economy of these regions.

Let's turn to the estimation of information indicators of unevenness. In economic research, one of the crucial tasks is the reduction of the initial size of the indicator space. This is due to the fact that regional economic systems have a complex multilevel structure with the verity of elements, subsystems and links. The initial number of variables that describe these systems is rather big and can't be taken into account without overloading of decision-making process. As a result, there is a problem of forming the system of the most informative, diagnostic indicators that allows to reduce the dimension of the original space without losing significant information. Methods of selecting representatives of groups allow to obtain a desirable system of indicators corresponding to above requirements. One of the most common methods for selecting representatives of groups is the "center of gravity" method. Let's discover its algorithm.

Step 1. In the first step of the algorithm, the matrix of initial data for each group of indicators is formed: $Y_1, Y_2, \ldots, Y_k, k$—number of groups:

$$Y_k = (y_{ij})_k, \tag{3}$$

$(y_{ij})_k$—value of the ith indicator in the jth period (or for the j-th object) in the kth qroup;
$i = \overline{1, m}, j = \overline{1, n}$;
m—number of indicators in the kth group;
n—number of periods (or objects) in the kth group.

Step 2. In the second step, the standardization procedure for each group is carried out: $Z_1, Z_2, \ldots, Z_k$.

Step 3. In the third step, the representatives that carry the information inherent to the particular group are chosen. The choice is held according to the following rules:

Rule 1. If the initial number of indicators in the group is greater than two, the sum of the distances of each indicator to other indicators of the group is calculated:

$$p_i = \sum_{j=1}^{m} p(z_i, z_j). \tag{4}$$

The representative is determined according to the following formula:

$$p_s = \min_i(p_i). \tag{5}$$

Rule 2. If the initial number of indicators in the group is equal two, the sum of the distances of each of the two indicators from the representatives selected by the previous rule, is calculated. The representative must have the greatest sum of the distances.

The applications of the proposed models for the regions of Ukraine are presented below.

Let's form the initial set of indicators of regional socio-economic unevenness. This set is formed according to the following structural groups of indicators: demography, labor market, education, household incomes and expenditures, health care, housing, offenses, culture, recreation and tourism, environmental factors, economic potential, financial potential, domestic trade, investment, agriculture, industry, transport and communication, organizational potential, foreign economic activity, innovative potential. Each of them characterizes a certain aspect of the regional development (see Table 3).

These indicators can better perform social and economic processes in the region and help to assess the unevenness level. They will be used to group regions according to their socio-economic development. The sources of the research are data publications of State Statistics Service of Ukraine. The initial data set covers all regions with the exception of the autonomous republic of Crimea and temporarily occupied territories for a time period 2010–2017 [56].

In order to qualitatively estimate the unevenness and calculate the corresponding coefficients, there is a need to reduce the number of initial indicators. There are two groups of such methods: methods of constructing integral indicators and methods for reducing the number of indicators.

We will determine group representatives based on the "center of gravity" method. Since indicators have different dimensions and units, there is a need for their balancing. It may be implemented through standardization. After that, the matrix of distances is calculated, which shows the degree of closeness for indicators within each group. The distances are calculated as Euclidean metrics. In this way, the selection of the representative indicators that have the most significant information inherent in the group is carried out. The result of the model is a set of representative indicators that describe the most important aspects of the research object. For initial groups in which the number of members was three or greater the representative indicator had the minimum sum of distances. In the case of two members the representative indicator had the maximum sum of distances from the appropriate representatives of other groups. As a result, the dimension of the initial set of indicators was reduced from 63 to 21. The final list of indicators is provided in the Table 4.

According to the scheme given on the Fig. 1 model of region grouping by the unevenness level must be realized within the next research step. This model is based on the cluster analysis techniques. The inputs of the model are homogenous groups of regions. The results are provided for the time period 2010–2016. For each year the homogenous groups of regions are obtained based on the initial set of indicators that consists of 63 items. The exception is 2017 period for which there is a lack of some indicators. That is why it is processed on the basis of the set of representative indicators.

It should be noted that the input data have a different dimension and unit of scale, so there is a need for data standardization.

We propose to apply Ward classification method and k-means method. The aim of the first method is to check the hypothesis about the number of clusters. The results of classification are given on the Fig. 2. X-axis presents regions, and the vertical axis—the distances between them.

As we see, throughout the whole period the tendency for regions to be divided into two clusters is clearly followed. The composition of clusters over the years is almost unchanged. Permanent objects in cluster 1 are Dnipropetrovsk, Donetsk, Lviv, Zaporozhye, Odesa and Kharkiv regions. These results indicate a high inertia conservation of regional development, which is caused by the previous accumulated potential. The distance at which clusters are united in one set is quite large. This

Table 3 Indicator system of regional socio-economic unevenness

Group	№	Indicator
Demography	x1	Population
	x2	Natural increase in population
	x3	Migration increase inpopulation
	x4	Number of marriages
	x5	Number of divorces
	x6	Average life expectancy at birth
Labor market	x7	Economically active population aged 15–70
	x8	Employed population aged 15–70
Education	x9	The coverage of preschool education institutions
	x10	Number of students of general education institutions per 10,000 population
	x11	Number of students of vocational technical schools per 10,000 population
	x12	Number of students of higher educational institutions per 10,000 population
Income of the population	x13	Household income
	x14	Average monthly nominal wages of full-time employees
	x15	Available income per capita
	x16	Percentage of population with per capita equivalent cash income a month below the subsistence minimum
Household income and expenses	x17	Distribution of households
	x18	Number of households with children
	x19	Structure of households expenditures
	x20	Structure of households incomes
Healthcare	x21	Availability of doctors of all directions
	x22	Availability of nurses
	x23	Availability of hospital beds
	x24	Number of visits to the clinic
	x25	Distribution of HIV-infected and AIDS patients
	x26	Incidence of active tuberculosis
Housing	x27	Housing level
	x28	Housing construction for 1000 people
Offense	x29	Number of detected crimes
	x30	Total number of victims of crime

(continued)

Table 3 (continued)

Group	№	Indicator
	x31	Number of people who died from crimes
	x32	Number of persons convicted by court sentences
Culture, recreation and tourism	x33	Library fund
	x34	Number of clubs
	x35	Collective accommodation
	x36	Number of persons placed in collective accommodation
	x37	Children's health and recreation facilities
	x38	The number of children who were in the summer in children's health and recreation facilities
	x39	Number of tourists served
Environment	x40	Reset contaminated return water in surface water objects
	x41	Emissions of pollutants into atmospheric air
	x42	Current costs for environmental protection
Economic potential	x43	Gross regional product per capita
Financial potential	x44	Cost-effectiveness of operating activities of enterprises
	x45	Consumer price indices
Domestic trade	x46	Retail turnover
Investments	x47	Capital investment per capita
Agriculture	x48	Agricultural product indices
	x49	Capital investment per capita
	x50	Labor productivity in agricultural enterprises
Industry	x51	Volume of sold industrial products (goods, services) per person
Transport and communications	x52	Transportation of goods by roads
	x53	Revenues from the provision of postal services and communications
	x54	Sending postal transfers and retirement benefits
	x55	Mobile subscribers
	x56	Subscribers to the Internet
Organizational capacity	x57	Number of legal entities
Foreign economic activity	x58	Coefficient of export coverage of import of goods
	x59	Total exports of services

(continued)

Table 3 (continued)

Group	№	Indicator
	x60	Total imports of services
	x61	Direct investment per person
Innovative potential	x62	Distribution of total expenditures on innovation activities
	x63	Number of innovative enterprises in industry

Table 4 List of representatives

Group	№	Indicator
Demography	x4	Number of marriages
Labor market	x7	Economically active population aged 15–70
Education	x11	Number of students of vocational technical schools per 10,000 population
Income of the population	x14	Average monthly nominal wages of full-time employees
Household income and expenses	x18	Number of households with children
Healthcare	x26	Incidence of active tuberculosis
Housing	x28	Housing construction for 1000 people
Offense	x29	Number of detected crimes
Culture, recreation and tourism	x36	Number of persons placed in collective accommodation
Environment	x40	Reset contaminated return water in surface water objects
Economic potential	x43	Gross regional product per capita
Financial potential	x44	Cost-effectiveness of operating activities of enterprises
Domestic trade	x46	Retail turnover
Investments	x47	Capital investment per capita
Agriculture	x50	Labor productivity in agricultural enterprises
Industry	x51	Volume of sold industrial products (goods, services) per person
Transport and communications	x53	Revenues from the provision of postal services and communications
	x55	Mobile subscribers
Organizational capacity	x57	Number of legal entities
Foreign economic activity	x60	Total imports of services
Innovative potential	x62	Distribution of total expenditures on innovation activities

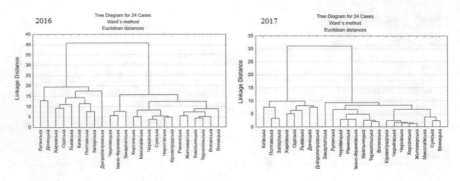

Fig. 2 Dendrograms of classification for time period 2016–2017

indicates a great sacristy between the objects of each cluster, that is, the unevenness of the indicator values is present.

Ward method Dendrogram showed not only a clear division into two clusters, but also a kind of sustainability of cluster structure. Thus, the unevenness of the socio-economic development of the regions was confirmed.

The dynamics of the transitions of regions between clusters by the Ward method is shown in Fig. 3. Let's investigate the obtained results.

The first class is characterized with the high level of socio-economic potential. It consists of Dnipropetrovsk, Donetsk, Zaporozhye, Kiev, Luhansk, Lviv, Odessa and Kharkiv regions. Poltava and Mykolaiv regions have an unstable nature of development. All other objects belong to the second class with low socio-economic potential.

Let's discuss the results of k-means method. Since the previous method determine the optimal cluster structure of two elements, we will also divide the objects into 2 groups. Euclidean distances between received clusters for each year are presented on the Fig. 4. As we can see, the distance between clusters gradually decreased in the period 2010–2013. This may indicate both the rise of weak regions and the decline in the level of development of strong ones. However, starting in 2015, the reverse process was held, and the distance between clusters increased. It indicates an aggravation of the situation of unevenness among the regions.

The mean values of indicators within each cluster for the time period 2016–2017 are shown on the Fig. 5. On the charts, you can see that the mean values of the two clusters intersect. It can also mean instability of the unevenness level.

According to the mean cluster values, the following preliminary conclusions can be drawn. The first cluster is characterized by high socio-economic potential of regional development. The second cluster includes the regions of the country with a low potential. According to the charts, you can see their breaks in values when splitting into clusters.

Consider the mean values in more detail. For example, several indicators have very close or identical values for both clusters. Among these ones are: x7—Economically active population aged 15–70; x9—The coverage of preschool education

№	Region	2010	2011	2012	2013	2014	2015	2016	2017	Class of stability
1	Vinnitska	2	2	2	2	2	2	2	2	I
2	Volinska	2	2	2	2	2	2	2	2	I
3	Dnipropetrovska	1	1	1	1	1	1	1	1	I
4	Donetska	1	1	1	1	1	1	1	1	I
5	Gitomirska	2	2	2	2	2	2	2	2	I
6	Zakarpatska	2	2	2	2	2	2	2	2	I
7	Zaporigska	2	1	2	1	1	1	1	1	II
8	Ivano-Frankovska	2	2	2	2	2	2	2	2	I
9	Kyivska	2	1	2	1	1	1	1	1	II
10	Kirovogradska	2	2	2	2	2	2	2	2	I
11	Luganska	2	1	2	1	2	2	2	2	II
12	Lvivska	2	1	2	1	1	1	1	1	II
13	Mykolaivska	2	2	2	2	2	2	2	2	I
14	Odeska	1	1	1	1	1	1	1	1	I
15	Poltavska	2	2	2	2	2	2	2	1	II
16	Rivnenska	2	2	2	2	2	2	2	2	I
17	Sumska	2	2	2	2	2	2	2	2	I
18	Ternopilska	2	2	2	2	2	2	2	2	I
19	Kharkivska	1	1	1	1	1	1	1	1	I
20	Khersonska	2	2	2	2	2	2	2	2	I
21	Khmelnitska	2	2	2	2	2	2	2	2	I
22	Cherkaska	2	2	2	2	2	2	2	2	I
23	Chernivetska	2	2	2	2	2	2	2	2	I
24	Chernigivska	2	2	2	2	2	2	2	2	I
The share of regions of the first cluster		17 %	33 %	17 %	33 %	29 %	29 %	29 %	33 %	

classification of region's states	1 – cluster with high social and economic development		I – stable class with high inertia	
	2 – cluster with low social and economic development		II – unstable class with low inertia	

		Characteristics of cluster	
		The total inertia of the regions	
	Low – II	5	20,83 %
	High – I	19	79,17 %
The total number of regions with high development potential		17	
Regions with low inertia in the second cluster		2	11,76 %
Regions with high inertia in the first cluster		15	88,24 %

Fig. 3 Dynamics of regions' distribution by clusters in 2010–2017

institutions; x23—Availability of hospital beds; x24—Number of visits to the clinic; x34—Number of clubs; x58—Number of clubs. These indicators may be equal for developed and underdeveloped regions. That is why they can't help in the procedure of cauterization and must be excluded from the research.

For the first cluster with a high level of socio-economic development, the following indicators are characterized by relatively low values: x2—Natural increase in population; x6—Average life expectancy at birth; x10—Number of students of general education institutions per 10,000 population; x11—Number of students of vocational technical schools per 10,000 population; x16—Percentage of population with per capita equivalent cash income a month below the subsistence minimum; x22—Availability of nurses; x28—Housing construction for 1000 people; x33—Library

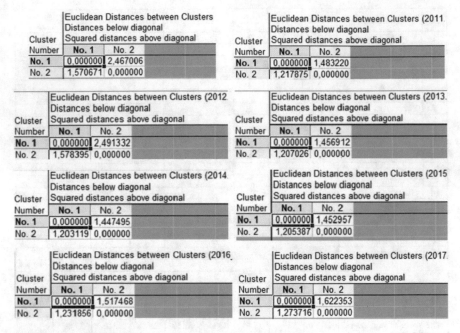

Euclidean Distances between Clusters Distances below diagonal				Euclidean Distances between Clusters (2011. Distances below diagonal		
Cluster Number	Squared distances above diagonal			Cluster Number	Squared distances above diagonal	
	No. 1	No. 2			**No. 1**	No. 2
No. 1	0,000000	2,467006		**No. 1**	0,000000	1,483220
No. 2	1,570671	0,000000		No. 2	1,217875	0,000000

Euclidean Distances between Clusters (2012 Distances below diagonal				Euclidean Distances between Clusters (2013. Distances below diagonal		
Cluster Number	Squared distances above diagonal			Cluster Number	Squared distances above diagonal	
	No. 1	No. 2			**No. 1**	No. 2
No. 1	0,000000	2,491332		**No. 1**	0,000000	1,456912
No. 2	1,578395	0,000000		No. 2	1,207026	0,000000

Euclidean Distances between Clusters (2014 Distances below diagonal				Euclidean Distances between Clusters (2015 Distances below diagonal		
Cluster Number	Squared distances above diagonal			Cluster Number	Squared distances above diagonal	
	No. 1	No. 2			**No. 1**	No. 2
No. 1	0,000000	1,447495		**No. 1**	0,000000	1,452957
No. 2	1,203119	0,000000		No. 2	1,205387	0,000000

Euclidean Distances between Clusters (2016 Distances below diagonal				Euclidean Distances between Clusters (2017. Distances below diagonal		
Cluster Number	Squared distances above diagonal			Cluster Number	Squared distances above diagonal	
	No. 1	No. 2			**No. 1**	No. 2
No. 1	0,000000	1,517468		**No. 1**	0,000000	1,622353
No. 2	1,231856	0,000000		No. 2	1,273716	0,000000

Fig. 4 Euclidean distances between clusters for time period 2010–2016

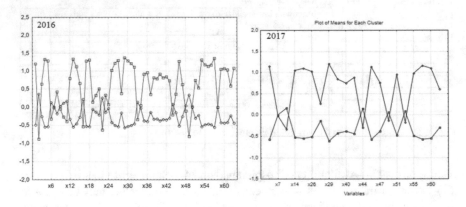

Fig. 5 Mean cluster values for time period 2016–2017

fund; x44—Cost-effectiveness of operating activities of enterprises; x49—Capital investment per capita; x50—Labor productivity in agricultural enterprises.

In other words, the developed regions are characterized by low fertility and, consequently, low life expectancy. But in the regions a smaller share of the population receives cash income below the country minimum because of higher wages. These regions are experiencing a lack of medical staff. There is also a growing need for new housing, because population migrates here from the second cluster. Labor productivity in agriculture is low, because these regions are less developed in this area. Accordingly, they have insufficient level of agricultural products production. All other indicators have high both positive and negative values.

They are: the level of crime, diseases, pollution of the environment; the level of GRP, wages; the quality of tourism and other sphere of regional development.

Let's proceed to the dispersion analysis results.

From the dispersion analysis for 2010, the worst classifiers are indicators x7, x33 and x44 (Economically active population aged 15–70, Library fund, Cost-effectiveness of operating activities of enterprises); the best classifiers are x46, x55–x57, x60 and x62 (Retail turnover, Mobile subscribers, Subscribers to the Internet, Number of legal entities, Total imports of services, Distribution of total expenditures on innovation activities). The largest contribution to the classification is made by x46 and x62 (Retail turnover and Distribution of total expenditures on innovation activities).

From the dispersion analysis for 2011, the worst classifiers are x7, x9 and x34 (Economically active population aged 15–70, The coverage of preschool education institutions, Number of clubs); the best classifiers are x46 and x57 (Retail turnover, Number of legal entities). The largest contribution to the classification is made by x57—Number of legal entities.

From the dispersion analysis for 2012, the worst classifiers are x9 and x23 (Number of legal entities, Availability of hospital beds), the best classifiers are x30, x46, x57 and x62 (Total number of victims of crime, Retail turnover, Number of legal entities, Distribution of total expenditures on innovation activities). The largest contribution to the classification is made by indicator x62.

The worst classifiers in 2013 are x7, x9 and x58 (Economically active population aged 15–70, The coverage of preschool education institutions, Coefficient of export coverage of import of goods), the best classifiers are x4, x46 and x57 (Number of marriages, Retail turnover, Number of legal entities). The largest contribution to the classification is made by x57—Number of legal entities.

The worst classifiers in 2014 are x9 and x24 (The coverage of preschool education institutions, Number of visits to the clinic); the best classifiers are x4, x5, x27, x46 and x53 (Number of marriages, Number of divorces, Housing level, Retail turnover, Revenues from the provision of postal services and communications). The largest contribution to the classification is made by x46—Retail turnover.

The situation was equal in 2015 and 2016 periods. The worst classifiers in 2015–2016 are x24, x34 and x58 (Number of visits to the clinic, Number of clubs, Coefficient of export coverage of import of goods); the best classifiers are x4, x5, x13, x17, x18, x27, x29, x30, x46 and x57 (Number of marriages, Number of divorces, Household income, Distribution of households, Number of households with children, Housing level, Number of detected crimes, Total number of victims of crime, Retail turnover, Number of legal entities). The largest contribution to the classification is made by x29—Number of detected crimes. The values of between group and within group dispersion of indicators in 2017 are shown in Fig. 6.

Fig. 6 Dispersion analysis
for 2017 period

Variable	Analysis of Variance (2017.sta)					
	Between SS	df	Within SS	df	F	signif. p
x4	15,77593	1	7,22407	22	48,04359	0,000001
x7	0,00340	1	22,99660	22	0,00325	0,955037
x11	1,27767	1	21,72233	22	1,29401	0,267549
x14	13,19731	1	9,80269	22	29,61848	0,000018
x18	14,43585	1	8,56415	22	37,08347	0,000004
x26	12,57366	1	10,42634	22	26,53094	0,000037
x28	0,87406	1	22,12594	22	0,86908	0,361328
x29	17,60162	1	5,39838	22	71,73185	0,000000
x36	8,60089	1	14,39911	22	13,14107	0,001498
x40	6,79330	1	16,20670	22	9,22165	0,006056
x43	9,45842	1	13,54158	22	15,36638	0,000733
x44	1,04097	1	21,95903	22	1,04292	0,318244
x46	15,52404	1	7,47597	22	45,68357	0,000001
x47	6,93166	1	16,06834	22	9,49049	0,005467
x50	0,21303	1	22,78697	22	0,20567	0,654621
x51	10,80028	1	12,19972	22	19,47637	0,000220
x53	0,38333	1	22,61667	22	0,37287	0,547700
x55	11,47452	1	11,52548	22	21,90271	0,000115
x57	16,01442	1	6,98558	22	50,43489	0,000000
x60	14,38847	1	8,61153	22	36,75842	0,000004
x62	4,34068	1	18,65932	22	5,11782	0,033900

The worst classifiers in 2017 are: x7 (Economically active population aged 15–70), the best classifiers are x29 and x57 (Number of detected crimes and Number of legal entities). The largest contribution to the classification is made by x29 (Number of detected crimes).

Thus, for the years 2010–2017, the region's membership in clusters was determined by the following indicators: the number of marriages, detected crimes, the turnover of retail trade and the number of legal entities. The x7 indicator is not statistically significant. The members of each cluster and their distances from the center of the corresponding cluster are shown in Fig. 7.

The figure shows that the distances between cluster centers and their objects are quite small. This indicates a fairly high level of density within each cluster, that is, the similarity objects within each cluster.

5 Discussion and Conclusion

According to the research results it was proved that system of regions of Ukraine consists of two relatively stable subsystems or clusters. These clusters are formed based on the unevenness level of socio-economic development of regions. The first cluster represents objects that have relatively high level of development. The following regions are included in it: Dnipropetrovsk, Donetsk, Zaporozhye, Kiev,

2010	Distance
Дніпропетровська	0,844269
Донецька	1,070462
Одеська	1,030389
Харківська	0,899625

2 кластер	Distance
Вінницька	0,609634
Волинська	0,573080
Житомирська	0,556439
Закарпатська	0,807317
Запорізька	0,804413
Івано-Франківська	0,892287
Київська	1,046970
Кіровоградська	0,677751
Луганська	1,050349
Львівська	1,062552
Миколаївська	0,690481
Полтавська	0,674551
Рівненська	0,682354
Сумська	0,495584
Тернопільська	0,701339
Херсонська	0,572194
Хмельницька	0,471265
Черкаська	0,650576
Чернівецька	0,676055
Чернігівська	0,767995

2011	Distance
Дніпропетровська	1,108868
Донецька	1,439814
Запорізька	0,786423
Київська	1,087522
Луганська	0,808697
Львівська	1,020868
Одеська	0,962444
Харківська	0,795035

2 кластер	Distance
Вінницька	0,557584
Волинська	0,491876
Житомирська	0,506973
Закарпатська	0,889809
Івано-Франківська	0,929146
Кіровоградська	0,591419
Миколаївська	0,772611
Полтавська	0,747143
Рівненська	0,628130
Сумська	0,493220
Тернопільська	0,645146
Херсонська	0,522320
Хмельницька	0,455448
Черкаська	0,568455
Чернівецька	0,657200
Чернігівська	0,686979

2012	Distance
Дніпропетровська	0,811197
Донецька	1,099947
Одеська	1,012032
Харківська	0,908051

2 кластер	Distance
Вінницька	0,497001
Волинська	0,533175
Житомирська	0,559637
Закарпатська	0,825588
Запорізька	0,818902
Івано-Франківська	0,921913
Київська	1,067466
Кіровоградська	0,649798
Луганська	0,948328
Львівська	1,064756
Миколаївська	0,743752
Полтавська	0,736428
Рівненська	0,689675
Сумська	0,572862
Тернопільська	0,746001
Херсонська	0,568369
Хмельницька	0,523512
Черкаська	0,561850
Чернівецька	0,714998
Чернігівська	0,676523

2013	Distance
Дніпропетровська	1,138757
Донецька	1,395421
Запорізька	0,813574
Київська	1,104559
Луганська	0,835269
Львівська	1,020141
Одеська	1,005816
Харківська	0,739841

2 кластер	Distance
Вінницька	0,571105
Волинська	0,500682
Житомирська	0,507278
Закарпатська	0,913110
Івано-Франківська	0,881381
Кіровоградська	0,646604
Миколаївська	0,777308
Полтавська	0,758395
Рівненська	0,634528
Сумська	0,528184
Тернопільська	0,700005
Херсонська	0,518631
Хмельницька	0,418841
Черкаська	0,513379
Чернівецька	0,686522
Чернігівська	0,642929

2014	Distance
Дніпропетровська	1,299455
Донецька	1,366508
Запорізька	0,753853
Київська	1,036224
Львівська	0,978912
Одеська	0,999048
Харківська	0,761413

2 кластер	Distance
Вінницька	0,672288
Волинська	0,519919
Житомирська	0,446174
Закарпатська	0,941157
Івано-Франківська	0,818091
Кіровоградська	0,628778
Луганська	1,183192
Миколаївська	0,673053
Полтавська	0,807854
Рівненська	0,634870
Сумська	0,541261
Тернопільська	0,678676
Херсонська	0,455311
Хмельницька	0,512037
Черкаська	0,479809
Чернівецька	0,679390
Чернігівська	0,627295

2015	Distance
Дніпропетровська	1,432431
Донецька	1,348203
Запорізька	0,754927
Київська	1,032421
Львівська	1,011225
Одеська	0,956043
Харківська	0,778039

2 кластер	Distance
Вінницька	0,717498
Волинська	0,521951
Житомирська	0,452781
Закарпатська	0,958395
Івано-Франківська	0,844239
Кіровоградська	0,605969
Луганська	1,110662
Миколаївська	0,641823
Полтавська	0,803412
Рівненська	0,591058
Сумська	0,514456
Тернопільська	0,630123
Херсонська	0,471510
Хмельницька	0,482914
Черкаська	0,464586
Чернівецька	0,640642
Чернігівська	0,611521

2016	Distance
Дніпропетровська	1,344874
Донецька	1,210280
Запорізька	0,766921
Київська	1,067486
Львівська	1,074257
Одеська	0,845114
Харківська	0,809622

2 кластер	Distance
Вінницька	0,767977
Волинська	0,655143
Житомирська	0,493060
Закарпатська	0,957671
Івано-Франківська	0,741137
Кіровоградська	0,557601
Луганська	1,097635
Миколаївська	0,668362
Полтавська	0,886319
Рівненська	0,524503
Сумська	0,598286
Тернопільська	0,648025
Херсонська	0,426106
Хмельницька	0,454191
Черкаська	0,498886
Чернівецька	0,694356
Чернігівська	0,550108

2017	Distance
Дніпропетровська	1,194350
Донецька	1,188550
Запорізька	0,740623
Київська	1,165934
Львівська	1,079884
Одеська	0,719958
Полтавська	0,937015
Харківська	0,848240

2 кластер	Distance
Вінницька	0,643892
Волинська	0,623386
Житомирська	0,308188
Закарпатська	1,259930
Івано-Франківська	0,448551
Кіровоградська	0,425768
Луганська	0,985617
Миколаївська	0,679942
Рівненська	0,564320
Сумська	0,607197
Тернопільська	0,646454
Херсонська	0,298809
Хмельницька	0,451818
Черкаська	0,403322
Чернівецька	0,490260
Чернігівська	0,463752

Fig. 7 Cluster members and distances for 2010–2017 period

Lviv, Odessa and Kharkiv regions. Vinnitsa, Volyn, Zhytomyr, Zakarpattia, Ivano-Frankivsk, Kirovograd, Lugansk, Nikolaev, Poltava, Rivne, Sumy, Ternopil, Kherson, Khmelnytsky, Cherkasy, Chernivetsy and Chernihiv regions have relatively low level of development and form the second cluster. It was proved that the majority of regions didn't move from one cluster to another during the study period 2010–2017.

Thus, the cluster structure in general is relatively stable. The socio-economic processes have the similar characteristics in regions from the same cluster, that is why the management decisions developed for one region may be implemented in other regions of the given cluster with minimum changes.

Despite this there are regions in both clusters which need special attention and additional research. For instance, Lugansk region belonged to highly developed regions at the beginning of the research period. However, the decline in its level of development since 2014 has led to a rapid transition of a powerful industrial region from the first cluster. At the same time, Lviv region has demonstrated a significant increase in development, which led to its inclusion into the first cluster. The Donetsk region holds its position in the first cluster due to the inertia of industrial production processes in spite of a rather large downturn. It is also necessary to draw attention to the Poltava region. It has moved to a group of highly developed regions in the last research period (2017). This indicates an increase in its development and the simultaneous downturn for the regions in the first cluster.

The proposed models should be implemented in the decision-making process and to became the foundation for efficient regional socio-economic policy.

Due to the fact that not all proposed in the conceptual scheme models got practical realization the future researches should be concerned to implementing them.

Acknowledgements We thank Anastasiia Didenko for assistance with calculation part of the research.

References

1. Ageyev D, Al-Ansari A, Qasim N (2015) Multi-period LTE RAN and services planning for operator profit maximization. In: Proceeding of the IEEE the experience of designing and application of CAD systems in microelectronics, pp 25–27. https://doi.org/10.1109/cadsm.2015.7230786
2. Ageyev D, Wehbe F (2013) Parametric synthesis of enterprise infocommunication systems using a multi-layer graph model. In: Proceeding of the IEEE 23rd international crimean conference microwave & telecommunication technology, pp 507–508
3. Ageyev D (2010) NGN network planning according to criterion of provider's maximum profit. In: Proceeding of the IEEE international conference on modern problems of radio engineering, telecommunications and computer science (TCSET-2010), pp 256–256
4. Al-Dulaimi A, Al-Dulaimi M, Ageyev D (2016) Realization of resource blocks allocation in LTE Downlink in the form of nonlinear optimization. In: Proceeding of the IEEE 13th international conference on modern problems of radio engineering, telecommunications and computer science (TCSET), pp 646–648. https://doi.org/10.1109/tcset.2016.7452140
5. Alehin E (2007) Regional economics and management. Gosudarstvennyiy universitet, Penza
6. Alexandrova A, Grishina E (2005) Uneven development of municipalities. J Voprosyi ekonomiki 8:9–105
7. Babeshko L (2006) Basics of econometric modeling. KomKniga, Moscow
8. Balta-Ozkan N, Watson T, Mocca E (2015) Spatially uneven development and low carbon transitions: insights from urban and regional planning. J Energy Policy 85:500–510. https://doi.org/10.1016/j.enpol.2015.05.013. Available from www.sciencedirect.com/science/article/pii/S0301421515002062, last accessed 2019/04/14
9. Baynev V, Pelih S, Baynev F (2007) Economy of the region. IVTs Minfina, Minsk

10. Bogashko OL (2006) Scientific and methodological principles of the region's economic development strategy. In-t M-va ekonomiki Ukrayini, Kiyiv
11. Bulgakova O (2003) Polarization of territorial development of meso-economic systems in the context of globalization. Rostov N/D, Rostov-on-Don
12. Bulman D (2016) Conclusion: a new political economy of uneven regional development. In: Incentivized development in China: leaders, governance, and growth in China's Counties. Cambridge University Press, Cambridge, pp 225–232. https://doi.org/10.1017/CBO9781316694497.007
13. Chagovets L, Prokopovych S, Chahovets V (2018) Estimation of structural-topological characteristics in information security system. In: Proceedings of international scientific-practical conference on problems of infocommunications. science and technology (PIC S&T), IEEE, Kharkiv, Ukraine, 9–12 Oct 2018, pp 265–270. https://doi.org/10.1109/INFOCOMMST.2018.8632165
14. Chepik A (2015) Investigation of the factors influencing the unevenness of economic development in the region. Izvestiya Sankt-Peterburgskogo gosudarstvennogo ekonomicheskogo universiteta 1:138–145
15. Cutrini E (2018) Economic integration, structural change and uneven development in the European Union. https://relocal.eu/wp-content/uploads/sites/8/2018/09/ERSA_Cutrini.pdf. Last accessed 2019/04/14
16. Danilenko A, Zimovets V, Sidenko V (2012) Risks and prospects of Ukraine's development in the period of post-crisis recovery. In-t ekonomiki ta prognozuv, Kyiv
17. Dobrynin I, Radivilova T, Maltseva N, Ageyev D (2018) Use of approaches to the methodology of factor analysis of information risks for the quantitative assessment of information risks based on the formation of cause-and-effect links. In: Proceedings of the 2018 international scientific-practical conference problems of infocommunications. science and technology (PIC S&T), Kharkiv, Ukraine, pp 229–232. https://doi.org/10.1109/infocommst.2018.8632022
18. Dubrov A, Mhitaryan V, Troshin L (2003) Mnogomernyie statisticheskie metodyi. Finansyi i statistika, Moscow
19. Egorshin O, Zosimov A (2001) Methods of multidimensional statistical analysis IZMN, Kyiv
20. Essays U (2019) The uneven distribution of economic activity across regions. Available from www.ukessays.com/essays/economics/the-uneven-distribution-of-economic-activity-across-regions-economics-essay.php?vref=1. Last accessed 2019/04/13
21. Friedmann J (2006) Regional development policy: a case of study Venezuela. The MIT Press Ltd
22. Galdin M (2004) Methodological approach to identifying the asymmetry of the socio-economic development of the region. GOU VPO Uralskiy gosudarstvennyiy ekonomicheskiy universitet, Omsk
23. Geets V, Shinkaruk L, Artomova T (2011) Structural changes and economic development of Ukraine. In-t ekonomiki ta prognozuv, Kyiv
24. Granberg A (2004) Basis of the regional economy. Vyisshaya shkola ekonomiki, Moscow
25. Gregus M, Kryvinska N (2015) Service orientation of enterprises—aspects, dimensions, technologies. Comenius University, Bratislava
26. Gritsay O, Ioffe G, Treyvish A (1991) Center and periphery in regional development. Nauka, Moscow
27. Grunig R (2005) Process-based strategic plannin. In: Grunig R, Kuhn R, 3rd edn. Springer-Verlag, Berlin
28. Hudson R (2013) Regional uneven development: ideas and approaches. In: To be presented at the Department of Geography, Harokopio University. https://radgeo.wordpress.com/κείμενα-άρθρα/ξενόγλωσσα/ray-hudson-regional-uneven-development-ideas-and-approaches. Last accessed 2019/04/13
29. International Indixes and Positioning of Ukraine. https://voxukraine.org/longreads/ratings/index.html#about. Last accessed 2018/12/11
30. Is Ukraine ready for the economy of the future: the country's place in world rankings? https://ukr.segodnya.ua/economics/enews/gotova-li-ukraina-k-ekonomike-budushchego-1105179.html. Last accessed 2019/01/10

31. Kaczor S, Kryvinska N (2013) It is all about services—fundamentals, drivers, and business models. J Serv Sci Res 5(2):125–154 Springer, The Society of Service Science
32. Kalinina V, Solovev V (2003) Introduction to multivariate statistical analysis. GUU, Moscow
33. Kazantsev S (2008) Assessment of the mutual position of the regions. J ekonomika i sotsiologiya 2:151–174
34. Ketova N, Ovchinnikov V (2006) Regional economy: universal educational economic dictionary. Feniks, Rostov-on-Don
35. Kirichenko L, Bulakh V, Radivilova T (2017) Fractal time series analysis of social network activities. In: Proceedings of the 2017 4th international scientific-practical conference problems of infocommunications. Science and Technology (PIC S&T), Kharkov, Ukraine, pp 456–459. https://doi.org/10.1109/infocommst.2017.8246438
36. Kirichenko L, Radivilova T, Tkachenko A (2019) Comparative analysis of noisy time series clustering. In: Proceedings of the 3rd international conference on computational linguistics and intelligent systems (COLINS-2019), vol I, Kharkiv, Ukraine, Apr 18–19, pp 184–196
37. Kirichenko L, Radivilova T, Zinkevich I (2018) Comparative analysis of conversion series forecasting in e-commerce tasks. In: Shakhovska N, Stepashko V (eds) Advances in intelligent systems and computing II. CSIT 2017. Advances in intelligent systems and computing, vol 689. Springer, Cham, pp 230–242. https://doi.org/10.1007/978-3-319-70581-1_16
38. Kirichenko L, Radivilova T, Zinkevich I (2017) Forecasting weakly correlated time series in tasks of electronic commerce. In: Proceedings of 2017 12th international scientific and technical conference on computer sciences and information technologies (CSIT), Lviv, pp 309–312. https://doi.org/10.1109/stc-csit.2017.8098793
39. Klebanova T et al (2015) Forecasting of social and economic processes. Simon Kuznets KhNEU, Kharkiv
40. Klebanova T, Guryanova L, Trunova T, Smirnova A (2009) Evaluation and analysis of the uneven development of the regions of Ukraine. Aktualnyie problemyi ekonomiki 8:162–168
41. Koychuev T (2014) About the uneven economic development of countries in the modern world. Obschestvo i ekonomika 6:5–12
42. Kravchenko T (2014) Adaptive of modeling of region economic development strategy. Can J Sci Educ Cult 2(6):894–900
43. Krugman P (2001) Increasing Returns and Economic Geography. J Polit Econ 99(3):483–499
44. Kryvinska N, Gregus M (2014) SOA and its business value in requirements, features, practices and methodologies. Comenius University, Bratislava
45. Kryvinska N (2012) Building consistent formal specification for the service enterprise agility foundation. J Serv Sci Res 4(2):235–269 Springer, The Society of Service Science
46. Maniv Z, Lutskiy I, Maniv S (2011) Regional economics. Magnoliya-2006, Lviv (in Ukrainian)
47. Mikhailova S, Moshkin N, Tsyrenov D, Sadykova E, Dagbaeva SD Spatial analysis of unevenness in the social-economic development of regional municipal units. Euro Res Stud J 20(2B):46–65. www.um.edu.mt/library/oar//handle/123456789/29263. Last accessed 2019/04/14
48. Molnar E, Molnar R, Kryvinska N, Gregus M (2014) Web intelligence in practice. J Serv Sci Res 6(1):149–172 Springer, The Society of Service Science
49. Pallares-Barbera M, Suau-Sanchez P, Le Heron R, Fromhold-Eisebith M (2012) Uneven development and regional challenges: introduction to the globalising economic spaces. In: Special Issue Urbani izziv 23(supplement 2):2–20. https://scholar.harvard.edu/files/montserrat-pallares-barbera/files/urbani-izziv-en-2012-23-supplement-2-000_introd.pdf. Last accessed 2019/04/13
50. Popov P (2010) Definition of socio-economic asymmetry of municipal organizations of the region. Sotsialno-ekonomicheskie yavleniya i protsessyi 5:85–88
51. Reshetilo V (2009) Synergy of the formation and development of regional economic systems. KhNGH, Kharkiv
52. Resolution of the Cabinet of Ministers of Ukraine of 20.05.2009 № 476 "On introduction of an assessment of interregional and intra-regional differentiation of socio-economic development of regions". http://zakon2.rada.gov.ua/laws/show/476–2009-p. Last accessed 2018/12/11

53. Sinchuk O et al (2009) Modeling of system characteristics in economics. MM, Kremenchuk
54. Skufina T, Baranov S (2017) The Phenomenon of Unevenness of socio-economic development of cities and districts in the Murmansk Oblast: specifics, trends, forecast, regulation. J Econ Soc Changes Facts Trends Forecast 10(5):66–82. https://doi.org/10.15838/esc.2017.5.53.5
55. Stasyuk O, Bevz I (2012) Integrated assessment of competitiveness of regions of Ukraine. Ekonomika i prognozuvannya 1:75–86
56. State Statistics Service of Ukraine. http://www.ukrstat.gov.ua. Last accessed 2018/12/05
57. Ukraine ranked the country with the smallest gap between the rich and the poor. https://www.epravda.com.ua/rus/news/2017/04/27/624325. Last accessed 2019/01/20
58. Umanets T (2007) Regional economic development of Ukraine: theoretical foundations of management, integral assessment, diagnostics. VIK, Donetsk
59. Vazhinskiy F, Kolomiets I (2004) The main methods of forecasting the socio-economic development of the region. Naukoviy visnik Natsionalnogo lisotehnichnogo universitetu Ukrayini 14(7):166–170
60. Zhalllo Ya, Pokrishka D, Belinska Ya (2011) Pislyakrizoviy rozvitok ekonomiki Ukrayini. Post-crisis development of the economy of Ukraine. NISD, Kyiv (in Ukrainian)
61. Zubarevich N (2008) Socio-economic development of regions: myths and realities of leveling. J SPERO 9:7–22

Formal Modeling of Decision-Making Processes Under Transboundary Emergency Conditions

Olga Cherednichenko⊙, Olha Yanholenko⊙, Maryna Vovk⊙
and Vasyl Tkachenko⊙

Abstract This paper is dedicated to the problems of modeling of decision making processes in the case of emergency occurrence. In particular, the transboundary emergencies are considered which impose constraints on the data used for decision making as well as on the process itself. The transboundary character of emergencies implies that the sources of data for environmental monitoring may be unavailable and the data gathered from the available sources, which are spatially distributed and belong to different state rescue services, may be incomplete, inaccurate or duplicate. The given study represents three types of problems related to emergency situations management: prevention of emergencies; immediate reaction on emergence that has just occurred; elimination of emergency consequences. The formal problem statement of these problems forms the basis for decision support system design. The effective management solution in an emergency situation is represented in the form of a sequence of actions that need to be taken under conditions of limited resources available for the elimination of emergency consequences. The paper represents the rules that allow the transformation of conditions of the formal decision-making problem statement into knowledge used by the environment monitoring system.

Keywords Decision support system · Environmental monitoring · Transboundary emergency · Formal model · Knowledge base

O. Cherednichenko (✉) · O. Yanholenko · M. Vovk
Kharkiv Politechnic Institute, National Technical University, Kharkiv, Ukraine
e-mail: olha.cherednichenko@gmail.com

O. Yanholenko
e-mail: olga.yan26@gmail.com

M. Vovk
e-mail: marihavovk@gmail.com

V. Tkachenko
Research Center of the Armed Forces of Ukraine "State Oceanarium", Odesa, Ukraine
e-mail: dbjyge@gmail.com

© Springer Nature Switzerland AG 2020   141
D. Ageyev et al. (eds.), *Data-Centric Business and Applications*,
Lecture Notes on Data Engineering and Communications Technologies 42,
https://doi.org/10.1007/978-3-030-35649-1_7

1 Introduction

Nowadays information technologies play a prominent role in the process of management and elimination of natural disasters and artificial accidents consequences [1]. Identification, combining and deploying of the related resources are the main activities which should be performed urgently after the disaster. The problem of the collection, processing, and analysis of data is one of the most pressing and unsolved problems nowadays [2–4]. The use of decision-making systems makes the process more efficient. According to numeral researches, the use of decision-making systems helps to solve these tasks.

Some researchers [5, 6] define the concept of support of the decision-making process as recognizing the situation for making optimal decisions using expert knowledge and methods of mathematical programming. Research of transboundary emergencies has its complex nature, also reflected in the decision-making process, which requires systematization and more structural approach. To support decision-making in the case of transboundary emergencies, many multi-criteria techniques, both isolated or integrated, should be used [7]. In order to make an optimal decision, alternative solutions should be formulated. Finding alternative solutions by using a decision support system (DSS) configured for a specific domain is a relevant task [5].

Information technology is used to implement DSS and provides decision-making process at all stages. The issue of comprehensive monitoring of the environment state often arises when it is necessary to determine the situation and make the right decisions [8–10]. The global environmental state is evaluated through the individual conditions expressed through different parameters. Therefore, it is necessary to track all environ-mental indicators simultaneously. The correlation structure in the data caused by interdependence between parameters provides relevant information which must be controlled. Thus, the complex structure of the data requires a multivariate control method to explore both the individual and interactive effects of the environmental parameters [11–13]. The main problems of collecting, processing and exchanging environmental information in the transboundary territories are described in [14]. The complexity of combined heterogeneous measurement data and flows of operational monitoring information make the problem even more comprehensive. Besides that, the mentioned papers describe many unresolved issues of information exchange between monitoring organizations of the country and edge states.

Specific electronic equipment for gathering and accounting data is used for estimating emissions and identifying the most hazardous substances [15–17]. Various personal devices are used to collect relevant information about the current state of the environment by the responsible agencies. Studies [18, 19] have shown that the impact of transboundary emergencies can be significant at national and local levels. Difficulties and complications in the estimation of hazardous level can restrain the possibility of effective decision-making to minimize these emissions. The collected information should be communicated quickly and uninterruptedly to the main office for coordinating activities and providing interaction between geographically separated locations.

As a rule, any territory already has some observation networks which belong to different services. Such separate devices and systems can monitor changes in the environment state, measure ecosystem health, and provide data for local, regional and national environmental of decision-making centers. Both current monitoring systems and traditional "manual" selection make an emphasis on data collection. Nowadays data are gathered using electronic measuring devices of remote monitoring in real time. The use of these measuring devices is performed via a connection to the base station or via telemetries network or landlines, cell phone network or other telemetry systems (Fig. 1).

Support of decision-making is based on the creation and use of structured information about the environment, and mechanisms of processing of these data to determine recommendations. The establishment of an internationally harmonized, integrated, and long-term operated environmental monitoring infrastructure is one of the major challenges of modern environmental research. The paper [20] suggests a service-oriented architecture of the real-time environmental monitoring system that is integrated with decision-support system of emergency recovery. It provides the

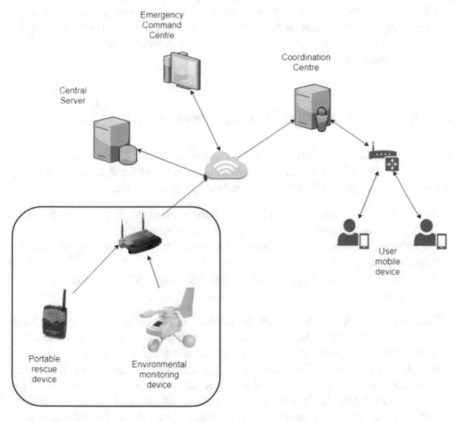

Fig. 1 Architecture of decision support system of environment monitoring [20]

mechanisms of communication between devices and interfaces to connect multiple devices into a single network. This approach allows providing the interface with compatible and incompatible devices, thereby making them interact efficiently. The benefit of the suggested architecture is that it allows deploying the emergency infrastructure in a quick manner and gathering the environmental information in a single data storage integrated with the decision support system. This in turn, increases the chances to eliminate the damage caused by an accident due to the well-judged and effective decisions. The challenge associated with the usage of the developed architecture is the volume of data that has to be processed and the processing itself. Since the collected data may be characterized as incomplete, inconsistent and inaccurate, the modern techniques of its filtering, cleaning, processing and estimating should be used [20].

As a result, all available data is collected and transferred to the main office, it should be used for decision-making by responsible authorities. The formal model's implementation for data processing in the DSS gives the possibility to increase the efficiency and creates the basis of information technology. The goal of the given paper is to represent a framework for modeling decision-making processes in the conditions of the transboundary emergency occurrence. The peculiarities of environmental emergencies themselves together with time and resources limitations require comprehensive research and application of modern modeling techniques to provide effective decision-making.

2   Modeling of Decision-Making Processes

The decision-making process depends on the developing stage of emergency. We can classify the tasks of decision according to existing resources and time for preparing the solution. It is proposed to distinguish three types of problems:

- the first type is a problem associated with the prevention of emergent situations (ES). It is usually resolved in a normal mode of operation;
- the second type is a problem associated with the immediate solution after the ES has occurred;
- the third type is a problem associated with the direct eliminating of ES consequences.

The task of the first type includes the steps of ES prevention and preparing long-term measures to reduce the risks associated with possible outcomes of ES. It is clear that the implementation of preventive measures requires a previous increase of material costs, but also leads to reducing the probability of ES occurrence. That is why these problems actually have risk management objectives [21, 22]. The task of the second type includes: the first, recognition and identification of ES; the second, the selection and appropriate immediate actions that can be performed with the assistance of available resources; the third, determining the necessary additional resources to prevent and overcome the effects of ES. The task of the third type presupposes the development of a system of measures that enables real-time overcome of the ES negative effects based on the different nature of ES and means to solve them.

Since the objective of the problem of the first type is to provide preventative actions, they should be formulated and elaborated in advance. This is usually done in the form of preliminary plans and programs. The tasks of the second and third types, by contrast, occur suddenly and require immediate response in real time, with all the time and resource constraints. In addition, it should be noted that to meet the challenges of the second type problem, only existing resources can be used and the time to involve them should be reduced, unlike the third type of tasks often require additional resources, which usually takes extra time.

In view of the above, the typical tasks of decision making in ES usually belong to the second and third types of tasks that are associated with ES. The processes of managing complex systems in the ES require a set of decisions on selection and implementation of activities performed sequentially in real time depending on the ecological system and the damaging effects on the environment [23–25].

Nowadays the risk of ES increases which is characterized by significant environmental and economic damage. Solving problems of the environmental information processing reduces risk and uncertainty of decision-making. The sources of initial information come from available sensors and software systems; they have different data format and data channels. In this regard, their processing feature is a need to harmonize heterogeneous data and their visualization in real time [23, 26]. In this paper, the method of collection and identification of information for decision support is suggested. It formalizes the processes of perception of environmental information through the use of finite predicates (Fig. 2).

The current state of environment in the ES is formally represented as the set of fulfilled conditions $P\tau$, and the set of events $T\tau$ performed at the real time τ. Let's apply these designations and record management processes in ES subject to the time interval $\Delta\eta$ as:

$$\left\{P_{\eta}^{(n)}, \{P\tau, T\tau\}, \tau \in \Delta\eta, P_{\eta}^{(M)}\right\} \tag{1}$$

where $P_{\eta}^{(n)}$ is a set of conditions fulfilled in the initial time moment of ES occurrence τ_0, $\tau_{0 \notin \Delta\eta}$;

P_{τ}, T_{τ} is a set of fulfilled conditions and a set of events performed in the current time moment τ;

$P_{\eta}^{(M)}$ is a set of conditions that determine the management goal during the ES, which provides the elimination of the consequences of ES;

η is an index of the specific implementation of management processes in ES, which arose for the object of automation in specific environmental conditions.

According to the subject (operator, manager, supervisor, etc.) and available resources that can be used for each pair $\left(P_{\eta}^{n}, P_{\eta}^{(M)}\right)$ that determines the start and the target state at the facility, there are various alternative management processes to

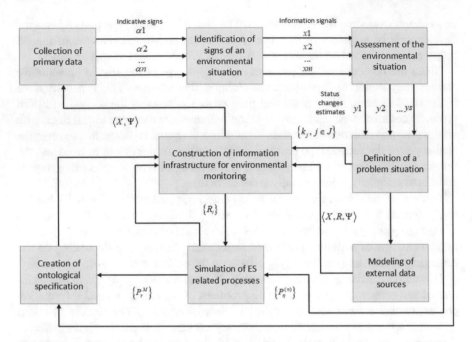

Fig. 2 Block diagram of the method of collection and identification of information for decision support in ES occurrence

overcome the ES. In order to identify and assess the quality of management decisions, it is necessary to define performance criteria φ, which can be represented, for example, in the form of minimizing the cost of resources or time spent to overcome the ES. Also, given the specific characteristics of each object, environmental criteria can also be used including criteria for minimization of undesirable effects of ES.

The set T_τ of activities performed in the current time τ is determined by the decision, which, in turn, depends on the set of conditions P_τ.

The process of overcoming the consequences of ES in the interval Δ_η is described as a sequence of sets

$$U_\eta = \left\{ T_\eta \middle| \eta \in \Delta_\eta \right\} \tag{2}$$

Obviously, for a given pair $\left(P_\eta^{(n)}, P_\eta^{(M)} \right)$ of sets on the range Δ_η and limits on the resources R_s, $s \in S_\eta$, mentioned criterion of efficiency of management $\varphi(U_p)$ clearly depends on the sequence of U_p of the form (2) of adopted and implemented solutions in order to eliminate ES consequences. Thus, the problem of effective management in ES can be formulated as the search for consequence

$$U_\eta^0 = \arg\left\{ \operatorname{opt}_{U_p \in U_\eta^*} \varphi(U_p) \right\} \tag{3}$$

under the restrictions:

$$P_\eta^n = P_\eta^{(n)};$$ (4)

$$P^{(M)} = P_\eta^{(M)};$$ (5)

$$R_s \leq R_{s\eta}, s \in S_\eta;$$ (6)

$$\widehat{R}_q \leq \widehat{R}_{q\eta}, q \in Q_\eta;$$ (7)

$$\tau_q \leq \tau_{q\eta}, q \in Q_\eta,$$ (8)

where

U_η^* is a family of sequences of the form (2) that are possible for a given object and specific negative environmental impacts on the time interval Δ_η;
$R_{s\eta}$ is the largest amount of resources of type s which can be used to overcome ES and need no preparation time;
$\widehat{R}_{q\eta}$ is the largest amount of resources of q type that can be used to deal with ES after training for a time not less than $q\tau_{q\eta}$;
S_η is a set of the ordered indices of resources types that can be used to overcome ES without spending time on preparation;
Q_η is a set of the ordered indices of resources types that can be used to deal with ES after training for the time not less than $\tau_{q\eta}$.

A system that is capable of performing the task of monitoring the environment must have the appropriate software, hardware, and information provision. There is a global problem controlling environmental parameters in transboundary areas for providing complete information and a clear plan of coordination in the case of emergency event. To estimate emissions and identify the most harmful factors, it is necessary to focus on data collection using electronic measuring devices and remote monitoring in real time.

During elimination of ES negative effects, there is a need in complete, accurate and timely information. Data can be collected by means of sensors and devices or by entering data through the user interface. To ensure that the operational center of ES effects elimination has all relevant information it is necessary to build a model of external data sources (information subsystem, devices, special equipment), which is the basis for the deployment of information infrastructure for environmental monitoring. In general, an external source is defined as a tuple $\langle K, Q, \Psi \rangle$ where K is the set of objects that define the functionality of the source, Q—set of relationships between objects, Ψ is the rules of data processing.

In real cases, ES occurrence is determined by the statement (1) based on monitoring data. Therefore, Eq. (1) defines a separate task of the second type, which is written as follows

$$P_{\tau 0} = P_\eta^{(n)}, \eta \in H, \tag{9}$$

where H is the set of ES types.

Lower index $P_\eta^{(n)}$ (9) is considered as a specific identifier of ES. Thus, the problem of overcoming ES $\varXi$ can be written as three types of problems:

(1) task of preventing ES ξ_1;
(2) task of immediate response ξ_2 at the time of ES occurrence;
(3) task of overcoming the consequences ES ξ_3

$$\varXi = \{\xi_1, \xi_2, \xi_3\} \tag{10}$$

The tasks ξ_1 are associated with planning preventive measures for ES and are qualitatively different from decision-making tasks directly during the ES. They were briefly discussed above. Tasks ξ_2 include:

– ES timely detection based on the Eq. (9);
– identification of ES that determines the type η of ES, which meets the conditions of (9);
– definition of the goal of ES overcome in the form of a set $P_\eta^{(M)}$;
– definition of the amount of necessary and available resources $\{R_{s\eta} | s \in S_\eta\}$, $\{\widehat{R}_{q\eta} | q \in Q_\eta\}$;
– implementation of urgent measures to protect environment and prevent the development of ES, and to prepare the necessary resources $R_s s \in S_\eta$.

The tasks ξ_3 involve finding and implementing action sequences (1) together, the best for a given situation $P_\eta^{(n)}$ and for initial efficiency criterion at a certain ES subject to the limitations on resources that provide achievement the target situation $P_\eta^{(M)}$, i.e. solving the problem (3)–(8). Described statement of decision-making related to emergency situations can be interpreted for domains with different objects. This makes it possible to consider the problem statement (2)–(8) as the default for widespread use.

Development of the typical problem statement of decision making after the occurrence of ES indicates the feasibility of understanding of this problem in the form of tasks ξ_2 to be solved immediately, and ξ_3 the solution of which is associated with the definition and consistent implementation of sets of measures (1) that best fit given performance criteria and conditions. Solving problems of group ξ_2 is first performed by continuously monitoring the current state of automation in order to discover and identify ES [using Eq. (9)] corresponding to the normal modes of a typical generalized model of ES [23]. After that, the process of problem solving is transmitted to the frame of decision making in the corresponding ES. Using the content of the slots and the associated procedures of this frame, the target management situation is determined in the form of sets of conditions $P_\eta^{(M)}$, available resources $\{R_{s\eta}\}$, $s \in S_\eta$ is a list of urgent measures to protect, prevention of the development of the ES and preparation of the necessary resources $\{\widehat{R}_{q\eta}\}$, $q \in Q_\eta$, as well as the implementation of these measures.

All solutions of these tasks must be received with the help of production rules stored in the Knowledge Base (KB) or in the data files in the database (DB) which describe specific ES. In contrast, the task ξ_3 for its solution requires specification and implementation in real-time the sequence of the sets of measures. The definition of measures to implement in the next moment of real time $T_{\tau+\delta}$ depends on the conditions that have been fulfilled in the implementation of measures T_τ in the previous time moment and the current state of the object and the environment. Obviously, the problem ξ_3 should be resolved on the basis of solutions of the problems ξ_2 and meet the frame at decision-making levels and activities of the typical generalized model of ES. Thus, the task ξ_3 matches statements (2)–(8) where the initial data η, $P_\eta^{(n)} P_\eta^{(M)}$, $\{R_{s\eta}\}$, $s \in S_\eta \{\hat{R}_{q\eta}\}$, $q \in Q_\eta$, is defined as the solution of problem ξ_2.

Tasks ξ_2 can be viewed as a special case of problems ξ_3. It should be noted that the sequence (1) is defined in tasks ξ_3 based on real-time decisions under the conditions of incomplete information. The information required to meet the challenges of management in ES is refined in the real-time interval Δ_η. An important feature of the problem ξ_3 is that the definition of the set of solutions (measures) T_τ may be based on large volumes of information (knowledge) in stock and at time τ, depending on the current state of automation and therefore is unknown. Thus, the task ξ_3 represents a new class of control problems and may be considered as a generalization of multistage stochastic control task [27, 28].

Analytic solution of the problem ξ_3 is obviously impossible, therefore, it should be formulated using principles which can create practical tools to make solutions for this task.

The principle of feedback provides an assessment of the current state of automation $\{P_\tau, T_\tau\}$ in the real time $\tau \in \Delta_\eta$ as a result of the implementation of decisions (actions) $\{T_\delta\}$ $\delta < \tau\delta \in \Delta_\eta$ to effectively clarify information about the current state of the object.

The principle of maximum parallelization of activities (transactions) of ES requires overcoming at every current moment $\tau \in \Delta_\eta$ to implement all measures T_τ for which certain conditions P_τ are fulfilled, including the preparation of resources $\{\hat{R}_{q\eta}\} q \in Q_\eta$. The implementation of this principle helps to minimize the time to overcome the effects of ES.

The principle of simulation of management processes is based on knowledge base by providing inference formation of the most effective solutions to overcome the ES through the use of large volumes of information accumulated in the intellectual core of ontological system in real time. The principle of optimization of alternative sequences of activities (1) is based on confidence and reliability factors. The implementation of this principle allows to find and implement the best way to overcome the ES specific criteria and the amount of knowledge in the system.

The principle of optimization operations (activities) T_τ, $\tau \in \Delta_\eta$ overcoming the ES provides the most efficient implementation, taking into account current information on the impending impact of the environment and condition of facilities.

The above principles can solve the problem (2)–(8) through the creation and implementation of intelligent information-analytical system (IAS). Using these principles IAS provides the following features:

- the use of feedback in terms of incomplete prior information;
- implementation of simulation processes in real-time ES;
- the benefits of expert systems in terms of collecting and compiling information, including intuitive representation of experts by applying fuzzy sets of coefficients and confidence indices;
- managing the large volumes of information in order to enhance solutions to overcome the ES by selecting the most plausible alternative sequences of actions (9).

The above results are the basis for the development of decision support algorithms to overcome the effects of ES on the basis of solving the problem (2)–(8). The development of ES is determined by numerous conditions and factors, therefore, to assess the decisions and appropriate actions, future consequences of these measures, it is necessary to use modeling of automation processes and control process, including a sequence (1) of measures in accordance with the decisions taken. Modeling tools can determine for a given initial situation $P^{(n)}$ the most effective strategy to achieve the goal set $P^{(M)}$ by management in accordance with the criteria.

A variety of processes take place in complex automation objects, hence there is a need to use different types of models and modeling techniques: analytical models; simulation modeling; graph models, including network models, such as Petri nets; physical models, including analog computing devices; full-scale models in the form of existing layouts, etc. [29–31].

The main modeling processes associated with ES is the integration of methods and models for comprehensive assessment of the consequences of decisions and the development of ES in the future. These estimates form the basis for selecting the best alternative solutions that optimize process control in ES. The basis for the integration of various models and processes in ES and mathematical tools are proposed in [32].

Process modeling on the basis of the generalized device model has a sequence of pairs

$$\left\{ P_\tau^M, T_\tau^M \right\}, \tau \in \Delta_{\eta M},$$

where $\Delta_{\eta M}$ is model time interval;

T_τ^M is recommended measures or software implementation models that have to be launched in the time $\tau \in \Delta_{\eta M}$ immediately after the set of conditions P_τ^M is fulfilled.

The results of simulation processes T_τ^M are considered as a new set of conditions $P_{\tau+\delta_r}$, where $\tau + \delta_r$ is the end of process modeling using the r-th model running at the time τ.

Thus, it can be considered as an integrated model that combines process modeling using inference on the basis of production rules and a variety of methods and models that describe processes in ES. An important feature is the integrated simulation model of decision-making and its implementation processes in facilities management, and the impact on these processes from the environment. In connection with this, it is advised to record the integrated model at each time $\tau \in \Delta_{\eta M}$ as a model of decision-making:

$$\{P_\tau^M, T_\tau^{MP}\}, \tau \in \Delta_{\eta M} \tag{11}$$

– models of processes in objects automation:

$$- \{P_\tau^M, T_\tau^{MO}\}, \tau \in \Delta_{\eta M} \tag{12}$$

– models of the effects of the environment as a series of conditions:

$$- \{P_\tau^{MC}\}, \tau \in \Delta_{\eta M} \tag{13}$$

The connection between model (11) and models (12)–(13) can be described through relations:

$$P_\tau^M = P_\tau^{MO} \bigcup P_\tau^{MC}, \tau \in \Delta_{\eta M} \tag{14}$$

$$T_\tau^M = T_\tau^{MP} \bigcup T_\tau^{MO}, \tau \in \Delta_{\eta M} \tag{15}$$

where P_τ^{MO} is the set of conditions fulfilled at time moment τ describing the current state of the control object;

P_τ^{MC} is the set of conditions fulfilled at time moment τ describing the impact of the current environment;

T_τ^{MP} is the set of operations (actions) that affect P_τ^{MO} and meet accepted at the time τ solutions using decision support system (based on ontological systems);

T_τ^{MO} is the set of operations (measures) and modeling tools or processes in the object, which started in time $\tau \in \Delta_{\eta M}$ and influence P_τ^{MO}.

Described methodology for modeling processes associated with ES, in addition to the benefits referred above, together with the general model (11) allows the use of different modeling approaches and decision-making methods with large amounts of information in order to improve decisions. Based on the above methodology, source data for modeling processes associated with ES is:

– the set of the conditions $P_\tau^{M(\Pi)}$ fulfilled in the initial situation;
– the time interval $\Delta_{\eta M}$ on which the problem of modeling is considered;
– the set of conditions ρ^{MO} that describe possible current state of automation (management) in the interval of modeling $\Delta_{\eta M}$, i.e. for each of the possible sequences $\rho_\tau^{MO}, \tau \in \Delta_{\eta M}$ the following relation takes place:

$$\left[U_{\tau \in \Delta_{\eta M}} P_{\eta r}^{MO} \right] \subseteq P^{MO}; \tag{16}$$

– a set of conditions $P_\eta^{M(M)}$ which determine the purpose of management in the ES;

– the set of production rules (5) representing the content of the active part of the ontological system:

$$P_{qu} \rightarrow \left\{ t^M_{u\delta}, P^*_u \right\} \tag{17}$$

where P_{qu} is the set of antecedent conditions, the implementation of which is necessary and sufficient for activities or launching models $t^M_{u\delta} \in T^{MP}_\delta$ that provide the completion of the conditions set fulfillment

$$P^*_u, P_{qu} \subset \rho^{MO}, P^*_u \subset \rho^{MO} \tag{18}$$

– T is set of measures (operations) to overcome ES, or models that can be implemented as a part of the rules of production

$$T = \left\{ T^{MP}_\delta \right\}, \delta \in \Delta_{\eta M}, \tag{19}$$

where the set of conditions (13) are given by (predicted) in the interval of modeling $\Delta_{\eta M}$ to take into account the impact of the environment on the process of overcoming ES; criterion for assessing the processes of overcoming the ES, which determines the numerical estimation of each of the sequences of decision-making (2) on the interval $\Delta_{\eta M}$ obtained by the simulation.

Based on this, integrated model (11)–(13) can be implemented on a computer in the form of ontological system with activation process output to knowledge base that contributes to the production rules (17). In the simulation process decision is represented as a model (11), which is based on inference and rules of productions of the form (17). The process control in ES is presented by a sequence (2) and is influenced by the sequence of decisions (11) and models the impact of the environment (13).

3 Formal Modeling of the Rules of Environmental Monitoring System's Knowledge Base

The task of creating an information technology for monitoring and operational coordination in emergency situations of a transboundary nature arises through the development of a set of models for collecting and identifying information to support decision-making under conditions of incomplete information in real time. A system that is capable of performing environmental monitoring tasks must have appropriate software, hardware, and information support. There is a global problem of environmental monitoring in transboundary areas, since full information and a clear coordination plan in case of an emergency is impossible.

To estimate emissions and identify the most harmful substances, emphasis has been placed on data collection using real-time electronic remote sensing measuring devices. A variety of personal devices used at the places of emergency services work they collect current information about the current state of the environment. The collected information should be delivered quickly and unconditionally to the command centres for the coordination of activities and providing communication between different geographically separate subdivisions.

In order to transform the conditions formulated in the generalized problem statement of decision-making in ES, it is necessary to formulate meta-rules that represent knowledge in the environmental monitoring system. This knowledge deals with the levels of substances that are considered as dangerous, values of all ecosystem parameters and their domains of definition, etc. When using the production representation of the base meta-rules, the interface between the monitoring system DB and the operating system (OS) should be built on the basis of a set of meta-rules that reflect, on the one hand, the database [33, 34], which is filled by the monitoring system, and, on the other hand, the structure of the meta-rules base (MB) which is determined by the variables included in the predicate of the antecedents of the meta-rules. The following notation will be used: Ω is the set of all meta-rules, $\Omega = \{\omega_r\}$, $r = [1, v]$, where v is number of rules; Y^* is a set of variables that are part of the predicates of the antecedents of the rules of the MB product $Y^* = Y$; P is the set of predicates that are included in the antecedents and the consecrates of the set meta-rules Ω; P_r is the set of all predicates that are part of the antecedent $\{P_{rs}^{(a)}\}$, $S \in V_r^{(a)}$ and the conjecture $\{P_{rh}^{(k)}\}$, $h \in V_r^{(k)}$ rules ω_r. There $V_r^{(a)}$, $V_r^{(k)}$ the set of predicate indexes according to the antecedent and the consequent rule ω_r.

Definition 1 The rule $\omega_i^{(1)}$, $i = [v + 1, m]$ is called the meta-rule of the first kind (MFK), if the antecedent of this rule consists of a predicate of the form:

$$P_{rs}^{(a)}(|\tilde{y}_i(t_n) - \tilde{y}_i(t_{n-1})| \geq \delta_i), \tag{20}$$

where $\tilde{y}_i(t_n)$ is the value of the variable y_i, $y_i \in Y^*$ at the current moment of measurement t_n; δ_i is the threshold estimate for the increment y_i over the time interval $[t_n - t_{n-1}]$, and the consequent MFK $\omega_i^{(1)}$ contains the predicate of the form:

$$P_{ih}^{(k)}(y_i = \tilde{y}_i(t_n)) = 1, \tag{21}$$

which indicates the insertion of a new value $\tilde{y}_i(t_n)$ into the database at time t_n.

The threshold value for each variable δ_i relative to each variable y_i is determined in accordance with the permissible error of the measuring channel, as well as the value of this parameter, and is entered in the predicates of the antecedents of the corresponding MFK when establishing the interface between the monitoring system of the current state of the ecological system in the conditions of the emergency and the OS.

It is clear that for a particular OS, the number of MFK corresponds to the number of variables Y^*, which are included in the predicates of the antecedents of the meta-rules:

$$|\Omega^{(1)}| = |Y^*|, \Omega^{(1)} \subset \Omega, \tag{22}$$

As can be seen from (22), MFK $\omega_i^{(1)}$ determines the fact that the value of y_i, which is controlled by the monitoring system, is significant for the MB.

Definition 2 The reflection of the content of the monitoring database in the OS is the process of calculating the values of the predicates of the antecedents $\{P_{rs}^{(a)}\}$, $s \in v_r^{(a)}$, for each of the meta-rules $\omega_r \in \Omega$, which depend on the value of the variable $\tilde{y}_i(t_n)$, $i = [1, m]$, which is measured in real time.

Definition 3 The production rule $\omega_i^{(2)}$ is called meta-rule of the second-kind (MSK), if the antecedent of this rule is a predicate of the form $P_j^{(a)}(y_i = \tilde{y}_i(t_n)) = 1$ and coincides with the consequent of the corresponding MFK $\omega_i^{(1)}$ and the MSK consequent $\omega_i^{(2)}$ is determined by the condition of the form:

$$\forall p_g^{(k)}(y_i = \tilde{y}_i(t_n)) = 1, \tag{23}$$

where each predicate $p_g^{(k)}$ corresponds to a definite predicate of the antecedents of the metarules with MB depending on y_i, $g \in G_i$.

For a given MB any MFK $\omega_i^{(1)} \in \Omega^1$ must correspond exactly to $\omega_i^{(2)} \in \Omega^2$. As can be seen from the definitions (20–22), meta-rules $\omega_i^{(1)}$, $\omega_i^{(2)}$ can be combined in one generalized meta-practice.

Definition 4 A general rule ω_i^u is called a generalized meta-rule (GMR) if its antecedent coincides with the antecedent of the MFK $\omega_i^{(1)}$, and the consequent is with the consequent MSK $\omega_i^{(2)}$.

Thus, the above method of organizing the interaction between the database and the MB allows continuously display, almost without delay, in the KB the current environmental state in order to ensure the functioning of IAS in real time, since the display of the values of database variables in the MB is practically at the moments of each significant change in the values of the corresponding parameters.

To combine the elements of individual ontologies that are part of the OS and to create a single knowledge space on this basis, the development of a special method is required. In the process of OS functioning, the following subsets $\Omega = \{\omega_r\}$, $r = \overline{1, V}$ of MB are formed in the set of rules: $\Omega^{(\kappa)}$ is the rules of the conflict set; $\tilde{\Omega}^{\kappa}$ is the rules for the applicants for inclusion in the conflict set; $\Omega^{(B)}$ is the rules that are executed at the moment; Ω^3 is the rules of production which operations have been completed at the current moment; $\Omega^{(HB)}$ is the rules of production which operations are not executed at the current moment within the allowed time interval; $\Omega^{(БЛ)}$is the rules blocked for a certain time to prevent their re-activation.

The permissible duration of the operations determined by experts in the rules of the subset $\Omega^{(HB)}$ can be exceeded at this moment, in particular, due to a malfunction of the elements of the system. The composition of the elements of the aforementioned subsets is constantly changed in such a way that at any t_n moment in MB one can distinguish the following subsets:

$$\widehat{\Omega}_n = \Omega_n^{(K)} \cup \widetilde{\Omega}_n^K \cup \Omega_n^{(B)} \cup \Omega_n^{(3)} \cup \Omega_n^{(HB)} \cup \Omega_n^{(БЛ)}, \Omega_n^* = {}^{\Omega}\!/_{\widehat{\Omega}_n}, \quad (24)$$

The subset $\widehat{\Omega}_{1n}$ is appropriately called the active part of the OS at time t_n, and Ω_n^*—is the passive part of this component at the t_n moment of time. The subsets $\Omega_n^{(K)}$, $\widetilde{\Omega}_n^K$, $\Omega_n^{(B)}, \Omega_n^{(3)}$, $\Omega_n^{(HB)}, \Omega_n^{(БЛ)}$ and Ω_n^* characterize the status of each rule of the KB in the current t_n moment of operation of the OS and, thus, the IAS state as a whole.

Definition 5 The current state of the MB at the t_n the moment of the operation of the IAS is determined by the composition of subsets of the active part $\widehat{\Omega}_n$ MB, which depends on the current state of the ecological system and the available expert information.

The role of each of these subsets is to accelerate the process of finding the knowledge in the OS environment by reducing the number of goals of the rules that are considered at each instant of time, as well as identifying the set of deviations in the state of the ecological system (as a result of the formation of the set $\Omega_n^{(HB)}$). The subset of the conflict set $\Omega^{(\kappa)}$ gives the possibility, with the help of a conclusion in knowledge, to formulate solutions in the form of a sequence of operations that are triggered directly as a result of measuring at the current moment t_n of the variables characterizing the ecological system.

A subset of the rules of the contenders for inclusion in the conflict set $\widetilde{\Omega}^\kappa$ serves for the rapid formation of the conflict set $\Omega^{(\kappa)}$ at the current moment of measurement of the object's parameters by the monitoring system. The composition of the elements of the subset $\Omega^{(B)}$ of the rules that are executed is determined by all operations implemented on the object at the current time. The subset Ω^3, completed at the current time of operations, is allocated to $\Omega^{(B)}$ and can be used in the subsystem of explanations to determine the path of the formation of solutions.

The subset $\Omega^{(B)}$ contains in its composition a subset of $\Omega^{(HB)}$ of the rules that have not been executed, that is, those rules for which the corresponding operations did not end within the allowed time interval (given a priori by experts), for example, violations in system operation.

The result of measuring $y_i(t_n)$ by the monitoring system at the current time t_n of any parameter $y_i \in Y, i = \overline{1, n}$ of the object can generally cause a change in predicate values that depend on this parameter $P_m(y_{i_i}) \in P, m \in V_m$ and, consequently, changes in the antecedents $\{p^{(a)}\}$ and the consequents $\{p^k\}$ of the rules of the BM, which includes predicate data; here V_m is the set of predicate indexes that depend on y_i.

We formulate the conditions for the inclusion and exclusion of product rules for the above subsets of MB directly after measuring the parameters in the t_n-th moment of operation. Obviously, the composition of $\Omega^{(\kappa)}$ includes the rules of production, all predicates of antecedents which are satisfied, that is

$$\omega_p \in \Omega_n^{(K)} \left| \left(\forall p_x \in \left\{ p_{p\alpha}^a \right\} \right) \right| p_x = 1, \ \alpha \in \mathcal{V}_p^{(a)}, \quad (25)$$

where is $\mathcal{V}_p^{(a)}$ set of predicates of antecedent of rule ω_p.

In the conflict set $\Omega_n^{(K)}$ only the subset rules $\tilde{\Omega}_n^K$ can be included, since the condition of satisfaction of all the predicates of the antecedent (sufficient condition) contains the satisfaction condition at the given time t_n of any part of the multiplicity of antecedent predicates (necessary condition).

The subset of the contender rules $\tilde{\Omega}_n^K$ for inclusion in the conflict set $\tilde{\Omega}_n^K$ is formed from the condition that there is at least one predicate in the antecedents of these rules that changed its value from zero to one as a result of the last measurement at the moment of time t_n.

$$\omega_m \in \tilde{\Omega}_n^{(K)} \left| \left(\exists p_\beta(y_{i,...}) \in \left\{ p_{m\Gamma}^{(a)} \right\} \right) \wedge (p_\beta(y_{i,...}) = 1) \right|, \Gamma \in \mathcal{V}_m^{(a)}, \qquad (26)$$

where $\mathcal{V}_m^{(a)}$ is the set of predicate indexes of the antecedent rules ω_m.

If as a result of measuring at the current t_n value of the variable $y_i(t_n)$ all antecedent predicates are not satisfied, then the relevant product rules must be excluded from the subset $\tilde{\Omega}_n^{(K)}$ immediately after the measurement (before the next measurement). To the subset $\Omega_n^{(B)}$ of the rules that are executed at the current time t_n, the rules $\Omega_n^{(K)}$ are added only from the conflict set (in the order determined by the scheme and the strategy of output on knowledge), $\Omega_n^{(K)} \subset \Omega_n^{(B)}$; in $\Omega_n^{(B)}$ also contains all previously executed rules, for which the following holds true:

$$\omega_j \in \Omega_n^{(B)} \left| (P(\delta_i - \tau_j \geq 0) = 1) \wedge (\exists p_y \in \{p_\sigma^{(K)}\}) | p_y = 0 \right), \qquad (27)$$

$$\tau_j = t_n - t_j, \sigma \in \mathcal{V}_j^{(K)}, \qquad (28)$$

where $\delta_j \in \Psi$, $j = \overline{1V}$ is set the duration of the operation that corresponds to the rule ω_j;

t_j is startup time to execute the rule ω_j;

$\mathcal{V}_j^{(K)}$ is the set of predicate indexes of the consequent ω_j.

Completion of the rule with $\Omega_n^{(3)}$, $\Omega_n^{(3)} \subset \Omega_n^{(B)}$, for which the condition is satisfied:

$$\omega_I \in \Omega_n^{(3)} \left| \left(\forall P_\eta \in \left\{ P_{Ip}^{(K)} \right\} \right| P_\eta = 1 \right) \rho \in \mathcal{V}_I^{(K)}, \qquad (29)$$

where $\mathcal{V}_I^{(K)}$ is the set of consequent predicate indexes y in the rules ω_I, are removed from the subset $\Omega_n^{(B)}$ and are included in $\Omega_n^{(БЛ)}$, that is, the predicates of the consequent of these rules were performed according to the measurement of the object carried out after the execution of the operation.

Rules of type $\omega_\xi \in \Omega^3$ that are already executed, in order to prevent their reuse (if the antecedents of these rules were once satisfied) should be blocked by the expert specified or stipulated by the regulation time $\{\lambda_r\} = \Lambda$, $r = \overline{1, V1}$ so at t_n-th moment of time in $\Omega_n^{(БЛ)}$it is necessary to include rules from $\Omega_n^{(3)}$ and all previously executed rules for which:

$$\omega_\xi \in \Omega_n^{(\text{БЛ})} \big| \big(P(\lambda_\xi - \bar{\tau}_\xi \geq 0) = 1, \ \tau_{\xi=}t_u - \bar{t}_\xi, \tag{30}$$

where $\bar{t}_\xi$ is the moment of completion of the operation which corresponds to the rule ω_ξ.

A subset of $\Omega^{(\text{HB})}$ meta-rules that are not executed on an admissible time interval $\Omega_n^{(\text{B})}$, $\Omega_n^{(\text{HB})} \subset \Omega_n^{(\text{B})}$, it is expedient to allocate from a subset in order to use these rules to analyse the causes of violations in the work of the object.

By $\Omega_n^{(\text{HB})}$ rules with an unsatisfied consequent are included for which the implementation time δ_q at the current time t_n was exceeded, that is,

$$\omega_q \in \Omega_n^{(\text{HB})} \big| \big(P(\delta_q - \tau_q \geq 0) = 0\big) \wedge \Big(\exists p \in \big\{P_{q\theta}^{(\text{K})}\big\} \big| q = 0\Big), \tag{31}$$

$$\tau_q = t_n - t_q, \theta \in \mathcal{V}_q^{(\text{K})}, \tag{32}$$

where $\mathcal{V}_q^{(\text{K})}$ is the set of predicate indexes of the consequent rule ω_q;

Obviously, the rules $\widehat{\Omega}_n$ of the active part of the MB that do not satisfy at any moment t_n any of the conditions (25)–(32) are included in the passive part Ω_n^*

In order to ensure the normal operation of the OS, it is of fundamental importance that the time Δ_u for the formation of subsets of the active part $\widehat{\Omega}_n$ at any t_n-th moment of the system's operation was less than the interval between the measurements of the object parameters by the monitoring system:

$$\big[t_{n+1} - t_n\big]_{\min} > \Delta_n, \tag{33}$$

The process of forming the subsets of the active part is based on the principle of dynamic decomposition of the MB.

Definition 6 Dynamic decomposition of MB at time t_n is the process of formation of sets based on the transition of rules from one set to another within the framework:

$$\Omega_n^{(\text{K})} \cup \widetilde{\Omega}_n^{\text{K}} \cup \Omega_n^{(\text{B})} \cup \Omega_n^{(3)} \cup \Omega_n^{(\text{HB})} \cup \Omega_n^{(\text{БЛ})} \cup \Omega_n^* \cup \Omega, \tag{34}$$

under conditions (25)–(32) for the time that satisfies condition (34).

To illustrate the decomposition process, let's consider at the time t_{n-1} the current state of the MB in the OS, which includes ten rules $\{\omega_r\}\Omega, r = \overline{1, 10}$, of which two have the status of those executed $\Omega_n^{(\text{B})} = \{\omega_4\omega_8\}$. Suppose that at the t_{n-1} time moment monitoring system receives a new value $y_i(t_{n+1})$ of parameter $y_i \in Y$. The predicates of the precedents of the rules $\omega_1\omega_3$ depend on this parameter in such a way that as a result of changing the values of these predicates the antecedent rules ω_1 were satisfied, and in the antecedent rules ω_3 there were predicates that have a zero value.

Then the current time moment immediately after $(n + 1)$ time point, at the time $t_{n+1} + \Delta_{n+1}$, де $\Delta_{n+1} \ll t_{n+2} - t_{n+1}$ where (25)–(32) under the conditions (25)–(32) will be determined by a set of subsets:

$$\overline{\Omega}_{n+1}^{(K)} = \{\omega_1, \omega_3\}; \; \Omega_{n+1}^{(K)} = \{\omega_1\}; \; \omega_3 \in \Omega_{n+1}^{(K)},$$

$$\overline{\Omega}_{n+1}^{(B)} = \{\omega_1\}; \; \Omega_{n+1}^{(3)} = \{\omega_8\}; \; \Omega_{n-1}^{(БЛ)} = \{\omega_8\}; \; \Omega^3 \cap \Omega^{(БЛ)} \neq 0,$$

$$\Omega_{n+1}^{(HB)} = \{\omega_4\}; \; \Omega^{(HB)} \subseteq \Omega^{(B)},$$ (35)

$$\Omega_{n+1}^* = \{\omega_2, \omega_3, \omega_5, \omega_6, \omega_7, \omega_9, \omega_{10}\},$$

The formation of the above subsets takes place in a time interval $t_{n-1} < \tau \ll t_{n-2}$.

We will assume that all the predicates of the antecedents of the rules ω_9, ω_{10}, as well as the consequent ω_1 depend on the value $Y_j(t_{n+2})$ of the measurement parameter $y_j \in Y, j = i$, which changed at the measurement moment t_{n+2}. At the same time, the current state $t_{n-2} + \Delta_{n+2}$ де $\Delta_{n+2} \ll t_{n+3} - t_{n+2}$ KB at time t_n is determined by the following subsets:

$$\overline{\Omega}_{n+2}^{(K)} = \{\omega_9, \omega_{10}\}; \; \Omega_{n+2}^{(K)} = \{\omega_9, \omega_{10}\}; \Omega_{n+2}^{(B)} = \{\omega_9, \omega_{10}\},$$

$$\Omega_{n-2}^{(3)} = \Omega_{n+2}^{(БЛ)} = \{\omega_1\}; \; \Omega_{n+2}^{(HB)} = \{4\};$$ (36)

$$\Omega_{n+2}^* = \{\omega_2, \omega_3, \omega_5, \omega_6, \omega_7, \omega_8,\}$$

For the formation of the current state of the ecological system in the OS after measuring the values of the function variables at each time t_{n-1} and t_{n-2} it is enough to check only two rules ω_1, ω_3 and ω_9, ω_{10} correspondingly, that in five times less than the total number of rules in the MB.

In the case of the predicate of antecedent changes its value from 0 to 1 the calculation should be done at the only antecedent of candidate rules. It causes the reduction of the rules number. Therefore, the approach presented above provides to significantly reduce the number of meta-rules, which should be checked at every step of processing. Also, it gives a possibility to accelerate knowledge-based reasoning by using the conflict set of candidate rules.

4 Discussion

Complex and heterogeneous environment monitoring systems need to be used during the management of transboundary emergencies. The full environment state can be estimated at real-time based on interacting with different monitoring system equipment. Environment monitoring presumes to observe and check properties or quality attributes of natural system. Such approaches have been developed in the given work.

Today, the development of decision-support systems for management in emergencies attracts attention of many researchers and practitioners. Often, such systems consider some specific type of emergency or some particular task of emergencies control.

For example, in paper [35] authors describe the system of oil spill accidents management in the seas. The presented mathematical models allow to predict the oil transformation and movement at the sea surface within time which depends on waves, wind, other natural factors. The inputs of the forecasting model include the starting spill position and date, type of the oil, its volume and duration, simulation length and time step [35]. The models were tested on available data of Copernicus Marine Environment Monitoring Service. The developed system does not interact directly with the sensor system that provides all those data on sea environment state in the real time mode. So the questions of interoperability of the developed software components and data sources external systems are out of the authors' scope.

Another work [36] is devoted to creation of the decision support system for groundwater pollution emergencies management. The system relies on the data from the groundwater monitoring network that covers 64 observation wells in the study area of Ljubljansko polje, Slovenia. The network provides the developed system with the data on chemical tests of water samples from the considered territory. In this case again, the system doesn't support the real-time data transfer from the data sensors and analyzers to the decision module. However, the proposed hydrological model allows to predict groundwater movement enough precisely and it is helpful in cases of pollutions for emergency managers.

The problems of elimination of earthquake outcomes are considered in the paper [37]. Authors suggest the logistics management system that allows to distribute supplies after the emergency has happened. The developed system provides managers with the possibility to transfer materials and goods from the distributed point through the emergency center right to the disaster point. The input information for the developed system is the location of emergency supplies, the location of emergency point and the points on the electronic map where the roads are interrupted. All this source data is collected by user manually and when entered to the system is processed by geographical information system engine. So actually, the developed system does not interact with any directly data sources in real time mode, instead it relies on user's manual inputs.

Compared to the described decision support systems, the system of transboundary emergencies management presented in the given work deals with the real-time source data. The developed approach allows building the distributed monitoring system that combines various data sources which are actually data sensors, devices or any other systems that gather information on environmental indicators. The suggested models of decision-making are suitable for analysis of information on the indicators independently of their location (water, air, and ground). The benefit of this unified approach is that it can be used to combine data collected by other systems (analyzed above) since often an emergency has a multi-faceted character. For instance, earthquakes are often followed by pollution of water resources. So the presented approach can be used to unify the overall data on the emergency and its influence on the environment and to provide decision-making recommendations on the upper level.

5 Conclusions and Future Works

In the process of decision-making in case of emergencies, the following stages are distinguished: monitoring of the environment state; organizing of subsystems interaction as the operational deployment of information infrastructure; collection and processing of data for making operational decisions; liquidation of emergency consequences. The conceptual model is proposed as a basis for solving the environmental monitoring problem. The conceptual model formalizes the interaction between data sources, methods and means of data collection and environmental indicators. The ecological system is considered as a complex object, the state of which cannot be estimated explicitly. The typical model of the decision-making problem in the emergencies is developed, and the ontological specification of knowledge representation has been developed. It allowed constructing the intellectual system core and provides data processing in the search processes and analysis of precedents. The technology of cooperation in case of emergencies consists of the following steps: the definition of emergency classification features, the organizational structure formation of the emergency response staff, the information infrastructure organizing and the preparation of operational reports. The method of constructing a flexible information infrastructure applies the concept of "information as a service" and the device metamodel. This allowed reducing the deployment time of the information system and increasing the efficiency of information and analytical decision support systems in case of emergencies and the elimination of the emergency consequences.

The future work is oriented on deployment of the developed decision support system and its specific components into the infrastructure of emergency services and centers.

References

1. Yannikov I, Ponomareva D (2017) Modern systems for monitoring and forecasting emergencies. In: Innovative technologies in engineering: a collection of works of the VIII international scientific and practical conference 2017, pp 209–213
2. Sanginova O, Bondarenko S, Andriyuk V (2017) Computer-integrated system of monitoring and forecasting the quality of water bodies. Mater V Int Water Forum Water Res Climate 2:146–149
3. Wang H, Xu Z, Pedrycz W (2017) An overview on the roles of fuzzy set techniques in big data processing: trends, challenges and opportunities. Knowl-Based Syst 118:15–30
4. Zhukov A, Leonov P (2018) Problems of data collection for the application of the data mining methods in analyzing threshold levels of indicators of economic security. KnE Soc Sci Humanit 3(2):369–374
5. Power DJ, Sharda R, Burstein F (2015) Decision support systems. John Wiley & Sons, Ltd

6. Kryvinska N (2012) Building consistent formal specification for the service enterprise agility foundation. J Serv Sci Res 4(2):235–269 Springer, The Society of Service Science
7. Henrique dos Santos P, Neves SM, Sant'Anna DO, Henrique de Oliveira C, Carvalho HD (2018) The analytic hierarchy process supporting decision making for sustainable development: an overview of applications. J Cleaner Prod. https://doi.org/10.1016/j.jclepro.2018.11.270
8. King C, Thomas DSG (2014) Monitoring environmental change and degradation in the irrigated oases of the Northern Sahara. J Arid Environ 103:36–45
9. Thu HN, Wehn U (2016) Data sharing in international transboundary contexts: the Vietnamese perspective on data sharing in the Lower Mekong Basin. J Hydrol 536:351–364
10. Chettri N et al (2015) Long term environmental and socio-ecological monitoring in transboundary landscapes. International Centre for Integrated Mountain Development (ICIMOD), Kathmandu, Nepal
11. Mulema SA, García AC (2019) Monitoring of an aquatic environment in aquaculture using a MEWMA chart. Aquaculture. https://doi.org/10.1016/j.aquaculture.2019.01.019
12. Kaczor S, Kryvinska N (2013) It is all about services—fundamentals, drivers, and business models. Serv Sci Res 5(2):125–154 Springer, The Society of Service Science
13. Kryvinska N, Gregus M (2014) SOA and its business value in requirements, features, practices and methodologies. Comenius University, Bratislava
14. Razumov VV, Razumova NV, Alekseev OA (2017) Information requirements for terrestrial and space monitoring of precursors of natural and technogenic catastrophes on the border territories of Russia. Modern problems of remote sensing of the Earth from space 14(7):52–61
15. Transon J et al (2018) Survey of hyperspectral earth observation applications from space in the sentinel-2 context. Remote Sens 10(2):157
16. Frohn RC, Lopez RD (2017) Remote sensing for landscape ecology: new metric indicators: monitoring, modeling, and assessment of ecosystems. CRC Press
17. Bilitewski U, Turner A (2014) Biosensors in environmental monitoring. CRC Press
18. Plutalova TG (2017) Monitoring of transboundary territories. Water and ecological problems of Siberia and Central Asia, pp 80
19. Shabanov VV, Markin VN (2015) Conducting monitoring of water bodies in modern conditions. Publishing house of RGAU-MAHA
20. Tkachenko V, Cherednichenko O, Godlevskyi M (2018) The concept of device meta-model for real-time communication in the transboundary environment monitoring system. In: Problems of infocommunications. science and technology: PIC S & T. Proceedings of the IEEE. Kharkiv, Ukraine
21. Furems E (2008) Knowledge-based multi-attribute classification problems structuring. computational intelligence in decision and control. World Scientific Publisher, Singapore, pp 465–470
22. Molnar E, Molnar R, Kryvinska N, Gregus M (2014) Web intelligence in practice. J Serv Sci Res 6(1):149–172 Springer, The Society of Service Science
23. Tkachenko V, Cherednichenko O (2018) Information technologies of decision support in transboundary emergencies. aviation in the XXI century—safety in aviation and space technology. In: Proceeding of the eighth world congress. NAU, Kyiv, pp 94–98. Homepage. http://conference.nau.edu.ua/index.php/Congress/Congress2018/schedConf/presentations
24. Kirichenko L, Radivilova T, Zinkevich I (2017) Forecasting weakly correlated time series in tasks of electronic commerce. In: Proceedings of 2017 12th international scientific and technical conference on computer sciences and information technologies (CSIT), Lviv, pp 309–312. https://doi.org/10.1109/stc-csit.2017.8098793
25. Kirichenko L, Radivilova T, Zinkevich I (2018) Comparative analysis of conversion series forecasting in e-commerce tasks. In: Shakhovska N, Stepashko V (eds) Advances in intelligent systems and computing II. CSIT 2017. Advances in intelligent systems and computing, vol 689. Springer, Cham, pp 230–242. https://doi.org/10.1007/978-3-319-70581-1_16
26. Gregus M, Kryvinska N (2015) Service orientation of enterprises—aspects, dimensions, technologies. Comenius University, Bratislava

27. Dobrynin I, Radivilova T, Maltseva N, Ageyev D (2018) Use of approaches to the methodology of factor analysis of information risks for the quantitative assessment of information risks based on the formation of cause-and-effect links. In: Proceedings of the 2018 international scientific-practical conference problems of infocommunications. Science and technology (PIC S & T), Kharkiv, Ukraine, pp 229–232. https://doi.org/10.1109/infocommst.2018.8632022
28. Kirichenko L, Bulakh V, Radivilova T (2017) Fractal time series analysis of social network activities. In: Proceedings of the 2017 4th international scientific-practical conference problems of infocommunications. Science and technology (PIC S & T), Kharkov, Ukraine, pp 456–459. https://doi.org/10.1109/infocommst.2017.8246438
29. Ageyev D (2010) NGN network planning according to criterion of provider's maximum profit. In: Proceeding of the IEEE international conference on modern problems of radio engineering, telecommunications and computer science (TCSET-2010), pp 256–256
30. Ageyev D, Al-Ansari A, Qasim N (2015) Multi-Period LTE RAN and services planning for operator profit maximization. In: Proceeding of the IEEE the experience of designing and application of CAD systems in microelectronics, pp 25–27. https://doi.org/10.1109/cadsm.2015.7230786
31. Bede B (2013) Mathematics of fuzzy sets and fuzzy logic. Springer, Berlin
32. Kirichenko L, Radivilova T, Tkachenko A (2019) Comparative analysis of noisy time series clustering. In: Proceedings of the 3rd international conference on computational linguistics and intelligent systems (COLINS-2019), vol I. Kharkiv, Ukraine, Apr 18–19, pp 184-196
33. Ageyev D, Wehbe F (2013) Parametric synthesis of enterprise infocommunication systems using a multi-layer graph model. In: Proceeding of the IEEE 23rd international crimean conference microwave & telecommunication technology, pp 507–508
34. Al-Dulaimi A, Al-Dulaimi M, Ageyev D (2016) Realization of resource blocks allocation in LTE Downlink in the form of nonlinear optimization. In: Proceeding of the IEEE 13th international conference on modern problems of radio engineering, telecommunications and computer science (TCSET), pp 646–648. https://doi.org/10.1109/tcset.2016.7452140
35. Liubartseva S, Coppini G, Pinardi N, De Dominicis M, Lecci R, Turrisi G et al (2016) Decision support system for emergency management of oil spill accidents in the Mediterranean Sea. Nat Hazards Earth Syst Sci 16(8):2009–2020
36. Janža M (2014) A decision support system for emergency response to groundwater resource pollution in an urban area (Ljubljana, Slovenia). Environ Earth Sci 73(7):3763–3774
37. Qi C, Fang J, Sun L (2018) Implementation of emergency logistics distribution decision support system based on GIS. Cluster Comput. https://doi.org/10.1007/s10586-018-1983-8

Decision-Making Support System for Experts of Penal Law

Alexander Alexeyev and **Tetiana Solianyk**

Abstract This section substantiates the relevance of the task of developing an expert decision-making support system for criminal law specialists. The goals of developing an expert system, the tasks solved by the expert system are formulated, and the experts responsible for filling the system and the types of users are defined. For the effective implementation of the expert system, a special data set structure was proposed, based on the basic properties of the Criminal Code article. The data sets organized in this way allow them to be used in data mining in the future. The decision tree method was chosen as a method for extracting new knowledge from fact bases. Analyzed its capabilities and limitations. Two main ways of its construction are considered—manual and machine learning. The possible storage options for the obtained tree in the system are described. The developed system is built based on client-server architecture. The description of the main parts of the system and their functional capabilities is given. Various technologies for the implementation of this system as a separate application for existing platforms, and for the implementation of the cross-platform version are considered. The main modes of operation of the dialogue part of the system, obtaining a solution and interpreting the results obtained are described. An example of decision making using the developed system is given. The final part contains a discussion of the possibilities and limitations of the developed system, the positive and negative aspects of the methods and technologies used.

Keywords Decision-making support system · Decision tree method · Penal law

1 Introduction

Any society as model of the social relations reflects forms and extent of development of human relations, being in communication or occupied with joint activity. The state is a political form of the organization of society in a certain territory and has management personnel and coercions to which all population of the country submits.

A. Alexeyev · T. Solianyk (✉)
National Aerospace University "Kharkiv Aviation Institute", Kharkiv, Ukraine
e-mail: t.solianyk@khai.edu

© Springer Nature Switzerland AG 2020 163
D. Ageyev et al. (eds.), *Data-Centric Business and Applications*,
Lecture Notes on Data Engineering and Communications Technologies 42,
https://doi.org/10.1007/978-3-030-35649-1_8

The state is designed to provide law and order, establishment of the precepts of law governing the public relations and behavior of citizens, protection of the rights and freedoms of the person and citizen.

However, despite various levels of development of the state, offenses still exist in them and demand legal regulation.

Offense is the illegal behavior, guilty illegal socially dangerous act (action or inaction) contradicting requirements of precepts of law and made by the right capable person or persons.

It is accepted to bear responsibility according to norms for any offense administrative, labor, tax and other industries of the right. Separately from these offenses (offenses) crimes—the harmful actions repeating and gaining distribution, which caused response from society, are considered. The public danger is expressed in the formal criminal ban established by legislative (representative) bodies of the state.

In general, it is possible to tell that in society there are processes of criminalization and decriminalization of actions that go continuously.

Criminalization is a process of recognition of act criminal and fixing of its signs in the criminal law, establishments for it criminal responsibility.

Decriminalization is the return process connected with act recognition not criminal, an exception of its signs of the criminal law, cancellation of a criminal liability for its commission (perhaps, with establishment for it other types of responsibility, for example administrative) [1].

As a result changes, which are considered at implementation of criminal procedure—activities of bodies of preliminary investigation, the investigator, the prosecutor, the judge and court for disclosure of crimes, to exposure and a punishment of the guilty, and prevention of punishment of the innocent, are constantly made to the criminal legislation.

Crimes differ in severity, can be finished or incomplete, made by a group of persons, a group of persons by previous concert, organized group or the criminal organization.

During preliminary investigation the actus reus consisting of the following basic elements is defined:

Object of crime: public relations (actually object of crime) crime subject; the victim from crime.

Objective party of crime: socially dangerous act; criminal consequences; a causal relationship between act and the come consequences; place; time; way; situation; tool; means of crime execution.

Subject of crime: natural person; sanity; age of a criminal responsibility; signs of a special subject.

Subjective party of crime: wine; motive; purpose; emotional state.

Obligatory element of the indictment, which complete preliminary investigation, is the formulation of charge. The formulation of charge is the procedural document representing statement of the legislative formulation (model) of socially dangerous act enshrined in the criminal law taking into account specific circumstances of act, perfect this or that person and the instruction for point, a part, the article of the criminal code providing this act.

Need of the analysis and accounting of such set of the interconnected parameters for drawing up objective and fair charge on the one hand, and existence of the structured and accurately formulated counts on the other hand allow to speak about relevance of a problem of automation of Ukraine criminal code articles search, which would correspond to an actus reus. It will help to provide the reduction of time for decision-making by the responsible expert and decrease in probability of emergence of a mistake at a formulation of charge. The considered processes of preliminary investigation can be referred to difficult which are characterized in large volumes of the analyzed information, badly formalized procedures of a logical conclusion for decision-making and difficulty of use of traditional methods of multicriteria optimization.

It is connected with development of the decision-making support system for experts of penal law, which can be used at a stage of preliminary investigation. Therefore, it is expedient to develop the expert system of decision-making, which is an effective method of the solution of unstructured tasks in the field of complex systems and processes for experts of penal law.

Such systems represent a type of the computer information systems helping responsible with decision-making, at the solution of badly structured tasks by means of direct dialogue with the computer with use of data, knowledge, difficult mathematical and logiko-analytical models.

The following distinctive features characterize the decision-making support system:

- Orientation to the solution of badly structured (formalized) tasks characteristic, mainly, of high levels of management;
- Possibility of a combination of traditional methods of access and data processing to opportunities of mathematical models and methods of the solution of tasks on their basis;
- Orientation on the nonprofessional end user by means of use of a dialogue-operating mode;
- The high adaptability providing an opportunity to adapt to service conditions and solvable tasks.

At the same time, two DSS directions were formed:

1. Toolkit for developing recommendations, with the help of which they solve the following tasks:

   - formation of a variety of alternative solutions;
   - formation of a set of criteria for evaluating alternatives;
   - evaluation of alternatives by criteria (taking into account the importance of criteria and utility function);
   - selection of the best alternative, which is issued by the system as a recommendation.

2. Data preparation toolkit, through which they solve such tasks:

- database preparation (often multidimensional and containing complex relationships);
- organization of flexible and convenient access to the database through query generation tools;
- obtaining the results of queries in the form most convenient for subsequent analysis;
- use of powerful report generators.

Both tools provide a decision making process. However, the second toolkit is a preparatory stage for the first, because it only prepares the data, but does not convert them into the form of the specified selection model.

Usually, the decision-making support system includes: a data processing and storage subsystem; subsystem storage and use of models; user dialogue management subsystem.

Databases as part of a decision support system have a much larger set of data sources, including external sources as well as sources of non-computerized data. The primary source of data is databases of various information systems, reporting, information from the Internet, etc. It is necessary to use all the data that may affect the correctness of the decision-making.

Another feature is the ability to pre-compress data from multiple sources by co-processing them with aggregation and filtering procedures.

The data in the decision support system is of great importance. They can be used directly by the user or as source data for calculation using mathematical models. Along with providing access to the DSS data, it should provide the user access to decision-making models. This is achieved by introducing appropriate models into it and using a database in it as a mechanism for integrating models and communication between them.

2 Development of the Decision-Making Support System for Experts of Penal Law

The expert system is the intellectual system intended for the solution of tasks in some subject domain based on knowledge provided by experts, containing the knowledge base and supporting functions of justification and an explanation of decisions.

The variety of types of expert systems is rather high and their expanded classification is given in such works as. It should be noted that they can differ on solvable tasks; in a form of the result offered upon termination of consultation process; on technology which is the cornerstone of representation of knowledge, a conclusion of decisions and realization of dialogue with the user; on extent of integration; on execution, etc.

In progress in creation of expert system, there was a certain technology of their development consisting of six following stages: identification, conceptualization, formalization, realization, tests, trial operation.

Identification. At the stage the tasks, which are subject to the decision, are defined, the development purposes are revealed, experts and types of users are defined.

The purpose of system development—support of decision-making by experts of penal law for reduction of time for decision-making by the responsible expert and decrease in probability of emergence of a mistake at a formulation of charge.

Problems of an expert system:

1. Search of article of the Criminal Code corresponding to an actus reus.
2. Ensuring dialogue interaction of a system with the user at search implementation.
3. Interpretation of the received results in the form of the relevant article with the indication of its full contents.
4. Filling and editing knowledge base.

Conceptualization. At this stage, a meaningful analysis of the problem area has been carried out, the concepts used and their interrelations have been revealed, methods for solving problems have been identified.

This system is used for decision-making in the field of penal law.

The criminal legislation can exist as in the form of the uniform systematized (codified) compiled laws (code), and in the form of the separate acts containing concrete norms of penal law.

If the criminal legislation has the form of the uniform codified act, as a rule, in it are allocated the General and Special part. The general part includes norms in which the basic principles and other general provisions of penal law are established: the concepts and definitions connected with its main institutes, the list of types of punishments, etc. Norms of the General part, as a rule, have regulatory character: these are norms declarations, norms instructions, norms definitions; some of these norms have incentive or allowing character.

Special part includes the norms containing the description of concrete criminal actions and establishing concrete measures of punishment for them.

The main source of penal law of Ukraine is the Criminal Code of Ukraine—the uniform act regulating legal relationship which arise as a result of commission by the person of crime in particular establishes the volume of a criminal responsibility. The criminal code consists of 2 parts: The general and Special.

The general part contains 15 sections (Article 1–108). In it norms of a general meaning which define bases of a criminal responsibility are presented, provide a concept of crime, define types and limits of punishment for crimes, the bases of release from a criminal responsibility and serving sentence, etc.

Special part consists of 20 sections (Article 109–447). Each actus reus is described and the responsibility measure is established in it.

The competent analysis of subject domain is necessary for drawing up a good expert system. Specialists of this industry have to be engaged in the analysis of subject domain.

Experts have to define entities of subject domain and communication between them. It is necessary, for creation of model of an expert system. In this case creation of a decisions tree.

During the model design, it is necessary to pay attention not to structure of subject domain (in this case to the Criminal Code) as itself, but to distinguish to an essence of objects of subject domain entities from them.

Formalization. At this stage, tools are chosen and ways of representation of knowledge are defined, the basic concepts are formalized and ways of interpretation of knowledge are defined. ·

Data mining—the decision-making support process based on search in data of the hidden regularities. At the same time, the saved-up data are automatically generalized to information, which can be characterized as knowledge.

Methods for extracting new knowledge from fact bases used in data mining are quite different. These are statistical procedures, genetic algorithms, neural networks, decision trees, inductive logic programming, etc.

The complexity and diversity of data mining methods require the creation of specialized end-user tools for solving typical problems of analyzing information in specific areas. Since these tools are used as part of complex multifunctional decision support systems, they should be easily integrated into such systems.

Mean a set of the facts presented in compliance with the formulated purposes of their use by data.

One of options of storage of entities of subject domain is a data set. Structuring data in such look it is possible to use various options of machine learning, for the solution of problems of decision-making.

Data set Structure:

Structure of data—the program unit allowing to store and process a set of the same and/or logically connected data in computer facilities. The structure of data provides some set of the functions making its interface for addition, search, change and removal of data.

When structuring data in subject domain of the Criminal Code the major element is article. Each article belongs to a certain section of the Criminal Code.

Sections in itself bring logical division of articles on the basis of the victim. In case of UK of Ukraine sections of the Special Part, for example:

• Section 1

Crimes against bases of a homeland security of Ukraine;

• Section 10

Crimes against safety of production;

• Section 12

Crimes against public order and morality.

However, during developing the expert system, which is strongly leaning on structure of the Criminal Code, many concepts will be not clear for the inhabitant. Such system will be convenient only for experts of penal law.

When structuring a data set it is worth focusing on the main properties of article of the Criminal Code:

Table 1 Example of a fragment of such a data set

Name of article	Number of article	Punishment	Conspiracy	...
Illegal imprisonment or kidnapping	146	3 years imprisonment	+	...
Hostage taking	147	5–8 years imprisonment	–	...
...	...	...	...	...

- The identifier of article—number of article;
- Name of article;
- The punishment prescribed by article;
- The list of the actions coinciding with a subject of article.

An example of a fragment of such a data set is given in Table 1:

Organized data sets allow you to use them in data mining in the future.

Methods for extracting new knowledge from fact bases used in data mining are quite different. These are statistical procedures [2], genetic algorithms [3], neural networks, decision trees [4, 5], inductive logic programming [5], etc. This article describes an embodiment of the expert system using a decision tree.

A decision tree is a decision support tool used in machine learning, data analysis and statistics. The structure of the tree is a "leaves" and "branches". On the edges ("branches") of the decision tree, attributes are recorded on which the objective function depends, in the "leaves" the values of the objective function are recorded, and in the other nodes there are attributes on which the cases differ. To classify a new case, go down the tree to the leaf and give the appropriate value. Such decision trees are widely used in data mining. The goal is to create a model that predicts the value of the target variable based on several variables at the input. This method of data processing is often used because of the clarity of the response received.

Among the possible restrictions of this method, you can specify the following:

- Does not always find existing patterns due to the sequence of the analysis procedure;
- With a large number of rules, visibility is lost;
- It is not always possible to isolate the relationship "IF ... THAT ...".

Building a Decision Tree:

In general, tree nodes can be schematically depicted as follows (Fig. 1).

In this case, the node is a question and answer options. Each node has a unique identifier by which other nodes access it. A node must have more than one answer. The answer consists of the text of the answer, the article number that it is associated with and the identifier of the next node.

When one of the tree sheets is reached, the bypass of the decision tree is completed, and the algorithm produces the result in the form of a list of articles that correspond to the charge.

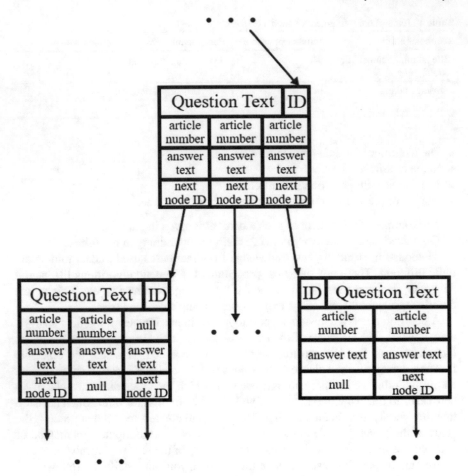

Fig. 1 Fragment of a typical decision tree structure

In this case, a simple search method is used to solve a decision problem using a traversal of a decision tree.

There are also many approaches to build a decision tree:

1. A decision tree can be built manually; for this, it is also worthwhile to involve experts in this subject area. In this case, experts can correctly evaluate entities and link them in such a way that the resulting decision tree will be easy to use by both specialists and ordinary people.

However, compiling a decision tree in a manual way leaves room for errors due to the human factor, this problem is deprived of the option of automatically building a decision tree.

2. Automatic decision tree building is a Data Mining task [6, 7]. For the task there are many algorithms, the most popular of them are:

- CART [8];
- ID3 [9];
- C4.5 [10].

However, this method has disadvantages.

To improve the usability of the expert system, the questions asked by the system should be written by a specialist in this field.

The algorithms for the automatic construction of a decision tree have a problem with retraining, as a result of which the tree can become too detailed, which will only confuse users.

To combat retraining, trees are cut, using pruning.

The advantage of using decision trees in decision making tasks is the simplicity of tree interpretation. Thanks to this, we can trace how the expert system came to a definite conclusion. In the case of such tasks as decision making in the area of the Criminal Code, this is a weighty advantage.

The data sets for the construction of decision trees practically do not need to be normalized, this is an advantage compared to such approaches in machine learning, such as:

- Artificial neural networks;
- Method of k-nearest neighbors [11].

Neural networks work only with numerical data types, and data domain such as the Criminal Code, it would be very difficult to normalize, the same problem is typical for the method of k-nearest neighbors. In addition, for data from an artificial neural network and a multidimensional representation of data from the k-nearest-neighbor method, it is very difficult to show the course of decision-making. When building decision trees data types do not play a big role.

Decision tree storage:

A tree, being a particular case of a directed graph, consists of nodes and arcs between them. To store the decision tree, you need to determine the structure and means of data storage.

One way to store a graph is a list of nodes that store links to other nodes.

You can store data in different formats; the most popular are JSON [12, 13], CSV [14], XML [15] and YAML [16]. The advantage of the JSON and YAML formats is good readability. There are also many serialization libraries for the listed formats [17], for most programming languages, which facilitates working with them.

In the case of a decision tree in the Penal Code, each node must have at least:

- Node ID
- Question text
- Answer options:
- Answer text
- Article number (may be empty)
- ID of the next node (can be empty)

Below is a variant of setting such a node in JSON format:

```
[ ...
{
    id: 50,
    question: 'Victim is adult?',
options: [
{
    text: 'Yes',
    article: null
    next: 51
},
{
    text: 'No',
    article: 148
    next: 51
},
]
},
...]
```

It is important to choose the right storage. For this, you can use both a simple file and database management systems (DBMS) [18].

In the case of a large amount of data, DBMS is the preferred option. You can use SQL-DBMS:

- Oracle [19]
- MySql [20]
- PostgreSql [21]
- MS SQL Server [22] etc.

   or NoSql DBMS:

- MongoDB [23]
- Neo4j [24]
- HyperGraphDB [25] etc.

Implementation. At this stage, the filling of the knowledge base is carried out. Since the basis of the expert system is knowledge, this stage is the most important and laborious. The process of acquiring knowledge is divided into the extraction of knowledge from an expert [26, 27], the organization of knowledge that ensures the effective operation of the system, and the presentation of knowledge in a manner understandable to the expert system.

Application Architecture:

The structure of the expert system is shown in Fig. 2.

   The developed expert system consists of the following main components: Knowledge Base Editor, Expert System Client Application, and Knowledge Base.

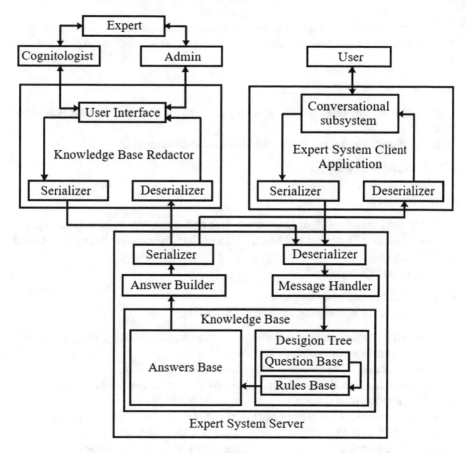

Fig. 2 Expert system's structure

The dialogue interaction subsystem is a combination of software tools that ensure the interaction of an expert system with various categories of users:

- Experts—describe the tasks, objects and processes for which the system is created;
- Cognitologist—describe tasks in the language of knowledge representation;
- Administrator—sets up the system and mechanisms for accessing databases;
- User—directly exploit the expert system.

Fact Base—needed to store the source and interim data of the current logical problem and interpret the rules in the process of logical inference.

The knowledge base includes a rule base and a question base.

The rule base is for the formalized description of logical problems in the language of knowledge representation of expert system. The base of questions—for replenishment, preservation and use of lists of possible questions for the selected logical problem.

Database—for the collection, preservation and issuance of permanent and operational information about the objects, events, facts, the results of solving problems.

Application architecture:

When designing any information technology, one of the most important stages is the choice of architecture.

In the case of an expert system, there is a clear division into:

- Knowledge Base [28];
- Solution system;
- User interface.

For the full functioning of the system, it is also necessary to have a knowledge base editor. The editor must modify, add, and delete the data in the target area and the decision tree.

Most expert systems tend to:

- A large amount of knowledge base;
- The importance of securing knowledge base data.

Based on these requirements, it makes sense to distinguish the client part and the knowledge base with the solution system.

This distinction corresponds to the client–server architecture.

Client—server is a computational or network architecture in which tasks or network load are distributed between service providers, called servers, and service customers, called clients [29]. The actual client and server are software. Typically, these programs are located on different computers and interact with each other through a computer network using network protocols, but they can also be located on one machine.

Client and server parts communicate using messages. Before editing, the messages are serialized into a convenient format for transfer, after receiving, the deserialization takes place back into the object.

The client part has an interactive subsystem that is convenient for communication with the user [30, 31]. When the message arrives at the server, the handler determines what action the client requests and performs the appropriate actions with the knowledge base. The knowledge base consists of a decision tree, which serves as a question base and rule base, and from the response base, which is used to build a response to the client. Regular users use the client application, while administrators and cognitive experts use the knowledge base editor in consultation with subject matter experts.

Consider the roles of the client and server for the expert system in the field of the Criminal Code:

Server Function:

The server in this case should store:

- The knowledge base of the subject area—the Criminal Code;
- Decision tree.

Data can be stored both in a file and in a DBMS.

The server also must communicate with the client in question-answer format. Registration in the system is optional, it gives more opportunities for keeping statistics.

When describing client communication with the server, it is important to choose the network protocol [32]. The most common are:

- UDP
- TCP
- HTTP

UDP protocol is unreliable and its use in this case is doubtful.

TCP is more convenient and reliable and is well suited for this task. However, the low-level capabilities of this protocol are not required in this case.

At the same time, the TCP-based HTTP protocol is more convenient and makes it possible to use the REST architectural style [33].

A good option is the WebSocket protocol, also based on TCP.

The server can be designed monolithically or distributed to various services.

In total, the server should, based on the data received from the client, return the result to it using the decision tree.

Client Function:

The client part of the application must provide a user interface to work with the expert system.

In general, the client must communicate with the server in a dialogue format. However, client capabilities are directly dependent on the server implementation.

When using TCP/UDP protocols, the client can be a mobile and/or desktop application. It should also use the TCP or UDP protocol, respectively.

Most of all opportunities are given by use of the HTTP protocol; at its use it is possible to implement clients in the form of:

- Web applications;
- Mobile application;
- Desktop application.

The WebSocket technology [34] is focused on interaction with web applications.

For the listed protocols, there are APIs for various social networks and instant messengers.

The knowledge base editor is also a client part of the system. His work can be organized in the following ways:

- The server application knows about the existence of the editor, receives messages from it and makes changes to the knowledge base.
- If all the elements of the knowledge base are stored in the DBMS, the editor can directly connect to it and make changes through requests.

Recommended Technologies:

The best option is the organization of communication between the client and the server using the HTTP protocol, it expands the choice of the client platform. Being a high-level protocol, it is quite easy to use and has a positive effect on the speed of development.

The development of mobile applications can be approached in two ways:

- Writing separate applications for each platform in their native language.

  - It is Java/Kotlin for Android [35];
  - It is Objective C/Swift for iOS [36].

- Writing cross-platform applications:

  - Flutter [37]—framework for language Dart
  - React Native [38]—framework for language JavaScript

Web application programming language JavaScript is used for writing. Such frameworks as:

- Vue JS
- React
- Angular 2

can help in writing web applications.

The server part can be implemented in almost any language. Here are a few options: PHP, Java, Kotlin, Python, Go, JavaScript и т.д.

The .NET platform has a stack of technologies for implementing this kind of systems [39]. Client applications can be implemented using:

- ASP.NET MVC Framework—framework for creating web applications;
- Xamarin—framework for creating cross-platform mobile applications;
- Windows Presentation Foundation—framework for building graphical applications for the Windows platform;
- LINQ и Entity Framework—libraries for working with DBMS.

For machine learning, the easiest way to use the language is Python, and framework:

- TensorFlow—computational back-end for neural machine learning [40];
- Keras—high-level API for creating models of neural networks. It is now part of the TensorFlow framework;
- Scikit Learn—Machine learning library;

The Expert System's Functioning:

The expert system, replacing the expert should provide a convenient interface for solving the task, and the result of its work should be well interpreted by the user [41, 42].

As a rule, the client communicates with such a system through a dialogue.

Dialog:

Client interaction with the expert system begins with a general question. The system also provides response options.

The question and answer choices correspond to the decision tree node.

In this case, the node, which were represented as JSON code above, corresponds to this system output:

Is the adult victim?

- Yes
- No

When you select one of the response options, the system remembers the answer and issues the next question. After answering the last question, the system returns the result of the decision tree.

Getting a Solution:

Because each answer question corresponds to a decision tree node, tree traversal depends on the user's choices.

The task of the expert system in the field of the Criminal Code is to compile a list of articles for the prosecution, in accordance with any illegal actions.

Before the beginning of the dialogue, an empty list is created, where the numbers of the articles of the Criminal Code will be added as the tree is walked around.

Each question has its own ID. As a rule, the answer option contains the ID of the next question. When selecting an answer, a question with the specified identifier and new answers is displayed.

In the absence of the next question (reaching a leaf of the tree)—the crawl ends and the system returns a list of articles.

Some answer choices also contain article numbers. Such a response. The meaning of such a response corresponds to a specific offense and when it is selected the article number is added to the list, which will be returned to the user upon completion of the expert system (Fig. 3).

A decision tree consists of nodes that store a question. Arrows personified by the variants of answers connect nodes with each other:

Interpretation of the Solution:

The result returned by the expert system must be well interpreted by both a specialist and a simple man in the street.

The result of a tree traversal is a list of articles that can be empty.

The user should receive not a list of values, but the display of articles containing:

- Article number;
- Article title;
- Description of the article;
- The estimated penalty for this article.

Fig. 3 Example of tree's
nodes connection

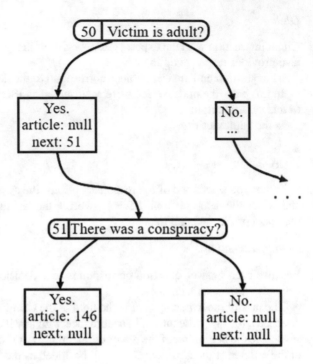

In the case of an empty list, the system should notify the user that no corpus delicti has been detected.

The information to be displayed is part of the knowledge base and should be stored on the server.

The searchable help system is a useful part of the client application.

3   Discussion

A year before this article was written, an expert system was created to help criminal law professionals. Understanding successful and erroneous decisions during the creation of an expert system formed the basis of this article.

Technology stack:

The expert system is a client—server application.

HTTP is used for communication between the client and the server. The data is sent as a JSON object serialized to a string, the received string is deserialized back to the object.

The server side is mostly implemented on the .NET platform. The server is written in the C # programming language, user information and the decision tree are stored in the Mircosoft SQL Server DBMS. The Criminal Code of Ukraine itself is stored in an XML file:

```
<?xml version="1.0" encoding="UTF-8" ?>
<criminal-code country="Ukraine" date="24.02.2018">
<part type="general" from="1" to="108">
<section id="1" name="    General provisions" from="1"
to="2"><article id="1" name=" Tasks of the Criminal Code
of Ukraine ">
```

1. The Criminal Code of Ukraine has as its task the legal provision of the protection of the rights and freedoms of a person and citizen, property, public order and public safety, the environment, the constitutional system of Ukraine from criminal encroachment, ensuring peace and human security, and the prevention of crime

```
. . .
</article>
</section>
```

The client part was implemented for the Android operating system. The application is written in JVM languages:

- Java8
- Kotlin

Both languages are native to the Android OS.

The interface of the client part looks like it is shown on the Fig. 4:

The first screen shows the decision process in the dialogue mode, the current question is displayed at the top of the screen. Below are the answers. The second form shows the result of the decision: a list of articles (title and content).

4 Conclusion

Guided by the rules and recommendations described above, an expert client-server architecture system based on the decision tree can be created.

The Criminal Code was chosen as the subject area, since the creation of an expert system in this area is an urgent task of our time.

The developed system belongs to the class of autonomous specialized static expert systems and has the following capabilities:

- contains a complete knowledge base for solving problems in the field of criminal law;
- correctly and consistently extracts information from the knowledge base;
- works directly in consultation with the user;
- presents knowledge, decision making and implementation of a dialogue with a user based on a decision tree.

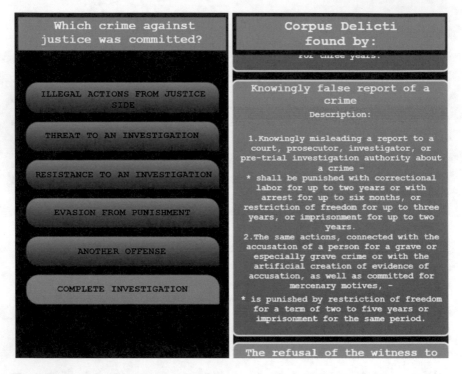

Fig. 4 The example of the client part interface

This expert system has such strengths and weaknesses:

Possibilities of the Developed Expert System:

The strong point is the high level of human interpretability of the process of obtaining the result by the expert system. This is an advantage of the decision tree method underlying the decision-making algorithm.

Thanks to the client-server architecture, the security of the knowledge base and the convenient possibility of its modification are ensured.

In case of changes in the domain, the expert system can be easily updated due to the tree structure. Simply insert, delete, or update a node in the decision tree.

The decision tree can be created manually or automatically using an algorithm.

The expert system returns a result that is understandable, to both a specialist and an average person (for a deeper understanding of the subject area, beginners are advised to add a help system).

Thanks to the server device, the client application can be implemented for most platforms.

Restrictions of the Developed Expert System:

Ease of use and correctness of the result mainly depend on the assessment of the subject area by experts. This minus is common to most expert systems.

The structure of the data set should provide for the possibility of building a decision tree. Accordingly, the experts who compose the data set must either have knowledge of decision trees or consult with machine learning experts.

The algorithms for the automatic construction of decision trees are far from perfect; the generated trees are often obtained too detailed and complicate the decision-making process. The resulting trees have to be reduced.

The generated tree reflects only the structure of the future decision tree (questions and answer choices should be rewritten for better understanding by humans).

References

1. Zakonodavstvo Ukrayiny (Legislation of Ukraine) The criminal code of Ukraine. Kiev, Ukraine (2019)
2. Bartroff J, Lai TL, Shih M-C (2013) Sequential experimentation in clinical trials: design and analysis. Springer
3. Alekseyev VY, Talanov VA (2005) Grafy. Modeli vychisleniy. Struktury dannykh (Graphs. Models of calculations. Data structures). Nizhny Novgorod, Russia
4. Kamiński B, Jakubczyk M, Szufel P (2018) A framework for sensitivity analysis of decision trees. Cent Eur J Oper Res 26:135. https://doi.org/10.1007/s10100-017-0479-6
5. Ray O, Broda K, Russo AM (2003) Hybrid abductive inductive learning. In: LNCS proceedings of the 13th international conference on inductive logic programming, vol 2835. Springer, Berlin, pp 311–328
6. Han K, Pei J, Micheline J (2011) Data mining: concepts and techniques (3rd ed). Morgan Kaufmann. ISBN 978-0-12-381479-1
7. Sadeghi J, Sadeghi S, Niaki S, Taghi A (2014) Optimizing a hybrid vendor-managed inventory and transportation problem with fuzzy demand: an improved particle swarm optimization algorithm. Inf Sci 272:126–144. https://doi.org/10.1016/j.ins.2014.02.075
8. Alpaydin E (2010) Introduction to machine learning. The MIT Press, London. ISBN 978-0-262-01243-0. Retrieved 4 Feb 2017
9. Taggart AJ, DeSimone A, Shih JS, Filloux ME, Fairbrother WG (2012) Large-scale mapping of branchpoints in human pre-mRNA transcripts in vivo. Nat Struct Mol Biol 19(7):719–721. https://doi.org/10.1038/nsmb.2327
10. Kuhn M, Johnson K (2013) Applied predictive modeling. Springer
11. Everitt BS, Landau S, Leese M, Stahl D (2011) Miscellaneous clustering methods. In: Cluster analysis, 5th edn. Wiley, Chichester
12. Crockford D (2019) How the JavaScript works. In: ECMA-404 The JSON data interchange standard. https://www.json.org/json-ru.html. Accessed 10 Apr 2019
13. Wieczorek S, Filipiak D, Filipowska A (2018) Semantic image-based profiling of users' interests with neural networks. Studies on the Semantic Web 36 (Emerging Topics in Semantic Technologies). https://doi.org/10.3233/978-1-61499-894-5-179
14. Shafranovich Y (2005) Common format and MIME type for CSV files. IETF. https://doi.org/10.17487/RFC4180
15. St. Laurent S (2003) Five years later, XML. O'Reilly XML Blog. O'Reilly Media. Retrieved 26 Oct 2010
16. YAML Ain't Markup Language (YAML) Version 1.2. YAML.org. Retrieved 29 May 2019
17. TC39 Proposal: Subsume JSON. ECMA TC39 committee. 22 May 2018
18. "database, n" (2013) OED Online. Oxford University Press
19. Giles D (2019) Oracle database 19c. Now available on Oracle Exadata. Retrieved 10 May 2019
20. MySQL 8.0 Release Notes. mysql.com. Retrieved 27 Apr 2019

21. PostgreSQL 11.4, 10.9, 9.6.14, 9.5.18, 9.4.23, and 12 Beta 2 Released!. PostgreSQL. The PostgreSQL Global Development Group (2019)
22. Download Microsoft SQL Server 2008 R2 (2011) Microsoft Evaluation Center. Microsoft Corporation
23. Witkowski W (2017) MongoDB shares rally 34% in first day of trading above elevated IPO price. MarketWatch. Dow Jones. Retrieved 26 Feb 2018
24. Philip R (2018) Simplicity wins: we're shifting to an open core licensing model for Neo4j enterprise edition. Retrieved 16 Jan 2019
25. Jaiswal G (2018) Comparative analysis of relational and graph databases. IOSR J Eng 3(8):25–27. https://doi.org/10.9790/3021-03822527. ISSN 2278-8719
26. Dobrynin I, Radivilova T, Maltseva N, Ageyev D (2018) Use of approaches to the methodology of factor analysis of information risks for the quantitative assessment of information risks based on the formation of cause-and-effect links. In: Proceedings of the 2018 international scientific-practical conference Problems of Infocommunications. Science and Technology (PIC S&T), Kharkiv, Ukraine, pp 229–232. https://doi.org/10.1109/infocommst.2018.8632022
27. Ageyev D (2010) NGN network planning according to criterion of provider's maximum profit. In: Proceeding of the IEEE international conference on modern problems of radio engineering, Telecommunications and Computer Science (TCSET-2010), pp 256–256
28. Green C, Luckham D, Balzer R, Cheatham T, Rich C (1986) Report on a knowledge-based software assistant. Readings in artificial intelligence and software engineering. Morgan Kaufmann, pp 377–428. Retrieved 1 Dec 2013
29. Otey M (2011) Is the cloud really just the return of mainframe computing? SQL Server Pro. Penton Media. Retrieved 1 Dec 2013
30. Kaczor S, Kryvinska N (2013) It is all about services—fundamentals, drivers, and business models. J Serv Sci Res 5(2):125–154 (The Society of Service Science, Springer)
31. Gregus M, Kryvinska N (2015) Service orientation of enterprises—aspects, dimensions, technologies. Comenius University in Bratislava
32. Network Protocols Handbook (2005) Javvin Technologies. ISBN 978-0-9740945-2-6
33. Richardson L, Amundsen M (2013) RESTful Web APIs. O'Reilly Media. ISBN 978-1-449-35806-8
34. Graham K (2011) IANA Uniform Resource Identifer (URI) Schemes. Internet Assigned Numbers Authority. Retrieved 10 Dec 2011
35. Android Language Breakdown (2017) Open Hub. Retrieved 15 Dec 2017
36. Juli C (2019) Apple releases iOS 12.3.1 with fixes for messages and VoLTE bugs. MacRumors. Retrieved 24 May 2019
37. Bracken C (2018) Release v0.0.6: Rev alpha branch version to 0.0.6, flutter 0.0.26 (#10010) flutter/flutter. GitHub
38. Windows Apps Team (2016) React Native on the Universal Windows Platform. blogs.windows.com. Retrieved 11 June 2016
39. Standard ECMA-335 (2012) Common Language Infrastructure (CLI), 6 edn. ecma-international.org. ECMA
40. An Open Source Machine Learning Framework for Everyone: tensorflow. Tensorflow (2019)
41. Ageyev D, Wehbe F (2013) Parametric synthesis of enterprise infocommunication systems using a multi-layer graph model. In: Proceeding of the IEEE 23rd international crimean conference microwave & telecommunication technology, pp 507–508
42. Al-Dulaimi A, Al-Dulaimi M, Ageyev D (2016) Realization of resource blocks allocation in LTE downlink in the form of nonlinear optimization. In: Proceeding of the IEEE 13th international conference on modern problems of radio engineering, Telecommunications and Computer Science (TCSET), pp 646–648. https://doi.org/10.1109/tcset.2016.7452140

Infocommunication Tasks of University Information System

Sergii Kavun⑩**, Olga Zavgorodnia**⑩ **and Anna Petrenko**⑩

Abstract Active usage of information technologies in every field of human activity nowadays is not a matter of choice. So any higher education institution as well as any other organization is forced to use information systems in order to be in the flow. So, it should be noted that real pace of changes in global education sector (especially its private part) is much quicker than stereotypes. Thus, higher education institutions eagerly try to meet these changes, though in most cases higher education institutions' infrastructure and information systems fail to do that. Consequently, there is a pressing scientific task of redefining infocommunication needs of higher education institutions through their information flows in order to identify actual modern requirements to the higher education institutions' information systems and infocommunication infrastructure. Those specific needs and requirements are identified on the basis of analysis of cutting-edge information technologies and trends that are used or expected to be used in education. As the result specific recommendations on infocommunication flows and university information systems integration are proposed.

Keywords Infocommunication flow · University information system · System integration and refactoring

1 Introduction

Active usage of information technologies in every field of human activity nowadays is not a matter of choice. So any higher education institution as well as any other organization is forced to use information systems in order to be in the flow.

S. Kavun (✉) · O. Zavgorodnia (✉) · A. Petrenko (✉)
Kharkiv University of Technology "STEP", Kharkiv 61010, Ukraine
e-mail: kavserg@gmail.com

O. Zavgorodnia
e-mail: olga.zavgorodnya@itstep.org

A. Petrenko
e-mail: anna.petrenko@itstep.org

© Springer Nature Switzerland AG 2020
D. Ageyev et al. (eds.), *Data-Centric Business and Applications*,
Lecture Notes on Data Engineering and Communications Technologies 42,
https://doi.org/10.1007/978-3-030-35649-1_9

According to the wide spread stereotype higher education institutions most likely are outdated in their IT infrastructure due to their tendency to the slow pace and non-risky development and scarcity of financial resources. This stereotype has some background underneath. Though the complexity of infrastructure used in higher education institutions is much often predefined by the historically set circumstances rather than actual infocommunication needs of higher education institutions (universities) and their financial capabilities.

During last decades, global education sector is developing not in a slow pace. One of the most promising part of it is computer-mediated education (conclusion based on [1]). According to a study held by Ambient Insight [2, p. 15], a computer-mediated education mostly falls into seven main groups: Self-paced e-learning courseware, Digital reference-ware, Collaboration-based learning systems, Simulation-based learning systems, Game-based learning systems, Cognitive learning systems, Mobile learning systems.

These digital educational products differ by multimedia content and technological innovations used, their popularity differ as well. So, some studies claim [2] that the global market for e-learning courses is currently over-saturated, so the main source of revenue for e-learning providers in the United States before 2021 should be: training simulations (17.0% of expected growth), game education systems (22.4% of expected growth), cognitive learning systems (11.0% of expected growth), mobile education systems (7.5% of expected growth). It is important to emphasize that such growth pace for traditional education services is outstandingly high.

The growth rate looks even more attractive if its global market capacity (in cash equivalents) is taken into account. It was estimated to be over $165 billion in 2015 according to the research [3] or $165,21 billion according to the other source [4]. The expected growth rate is also outstanding: it should fluctuate about the mark of 5% between 2016 and 2023 [3]. Another comparative study state that the growth pace should take about 7.5% of rising [4]. The expected volume growth forecasts are quite close in those estimations: global e-learning market is expected to reach $240 billion by 2023 according to [3] or $275,1 billion by 2022 according to [4].

Both researches seem to be close in their estimations and seem to be dubious at first glance due to such an optimistic forecast. The other research [5] seem to be much more cautious but it is also comparable to those estimations. The skepticism pops up mostly in the prospects of developing regions of the world. So most of the works claim that the unspeakable leader of e-learning implementation is Northern America, and this market is evidently saturated. Western European and Asian markets are next in its e-learning usage though it is much less volumetric and filled as well. The growth of Northern American and Western European markets is not expected at all (or even some reductions are possible). But the expectations of growth include Africa, Eastern Europe and the Russian Federation. The potential of Eastern European e-learning market claims to be attractive due to expected pace of growth, though in monetary terms its capacity is much lower than those of North America, Asia or Western Europe.

E-learning market capacity and growth rate are uniform all over the world but they still encourage higher education institutions introduce their services (educational products) to those markets. It highlights the need of universities to change their infocommunication infrastructure to meet the demands of global e-learning market.

Some universities have chosen the path of e-learning management systems introduction into their infocommunication infrastructure. The frontal statistical research on the point doesn't exist though it is possible to use the results of studies conducted under auspices of the European University Association [6, 7] and the EDUCAUSE [8]. Both organizations are influential on higher education market and have access to statistical data of vast majority of prominent universities. The European University Association includes as its members more than 800 universities in 48 European countries. It represents about 17 million students who are enrolled at the European University Association member universities. According to the surveys conducted under the auspices of the European University Association [6, 7], the vast majority of European universities have already implemented e-learning in the current educational process or are in the process of such implementation. The survey [6] was conducted on the basis of 249 higher education institutions in Western and Eastern Europe which are located in 36 countries. According to it [6], e-learning has been implemented by the vast majority of institutions surveyed (91%). Surveyed universities have introduced blended-learning systems, which involve integrating e-learning into a traditional system of teaching courses. The most interesting fact is that those universities state undeniable benefits of e-learning implementation but do not deploy it frontally. Only half of the institutions indicated that e-learning was implemented in the whole institution [6, 7], while only one third of institutions involved all or most students in e-learning. Consequently, it can be considered proven that almost all higher education institutions in Europe are involved in e-learning, but the degree of coverage of curricula by e-learning systems is not uniform.

The level of introduction of e-learning systems of surveyed universities into university information system might be different, but the infocommunication support by information technologies of some education resources are substantial [6]:

digital courseware (text books, reference materials, etc.) are supported throughout the institution (39% of universities) or in some faculties (46% of universities);
online repositories for educational material are supported throughout the institution (50% of universities) or in some faculties (31% of universities);
tools and management systems for content development and course management are supported throughout the institution (65% of universities) or in some faculties (23% of universities);
student portals are supported throughout the institution (76% of universities) or in some faculties (10% of universities).

Mentioned above information technologies are incorporated into university information system and successfully supported. Also it is important to mention the list of most widely provided IT-services for students by university information systems [6]:

online access to libraries (granted to all or most students of 93% of universities; to some students of 2% of universities);

access to computer rooms (provided to all or most students of 92% of universities; to some students of 4% of universities);

Wi-Fi (provided to all or most students of 92% of universities; to some students of 3% of universities);

university email accounts for all students (provided to all or most students of 88% of universities; to some students of 3% of universities);

campus licenses for software (provided to all or most students of 73% of universities; to some students of 11% of universities);

online study course catalogue (provided to all or most students of 71% of universities; to some students of 6% of universities);

social media to communicate with students or alumni (provided to all or most students of 63% of universities; to some students of 18% of universities);

personalized study portal (provided to all or most students of 50% of universities; to some students of 6% of universities);

electronic student portfolio (provided to all or most students of 29% of universities; to some students of 23% of universities);

online examinations (provided to all or most students of 24% of universities; to some students of 39% of universities).

The mentioned above IT-services are widely supported by University information systems and should be incorporated as must-have functionality for those which don't provide such serviced yet. As might be seen electronic student portfolio and online examinations are not so absolutely supported due to the cost and organization limits of higher education institutions.

Another source of reliable statistics on higher education industry is the EDU-CAUSE [8]—an international organization which comprises higher education professionals from over 2300 organizations. Member organizations of that community represent 45 countries, so even selective surveys under the auspices of EDUCAUSE are global, they can give a ground snapshot of the industry. The data from the elective study of e-learning covered 311 higher education institutions (EDUCAUSE members). According to it, more than 80% of institutions of higher education have at least several courses held with using e-learning systems, while more than half offer a significant number of electronic distance courses (disciplines).

Introduction of e-learning systems in prominent higher education institutions is a global trend which showed some important deviations from early expectations:

various e-learning models of education (mostly blended-learning models [9]) gradually enter all higher education institutions though not so comprehensive (as was anticipated);

the used infocommunication infrastructure for e-learning activities is still truncated (the functionality of used information system is poor, peer activities are scarcely supported, user activities are rarely and poorly monitored, customization of education serviced are not supported);

inclusion of e-learning systems into university information systems is held without proper integration (cross functionality mostly not supported, frequently the data are duplicated by those systems);

information security issues are faced with inadequate technology and qualification (much below the needed level) [10, 11];
modern, cutting-edge education technologies are not supported or rarely supported, even mobile education platforms (or systems) are rarely implemented.

Another promising but not so massive direction in infocommunication support of learning processes in universities is MOOCs (Massive Open Online Courses). Thus, according to [12, p. 14], the leader in the implementation of the MOOCs is North America, which is followed by Europe with a significant margin. According to the study of the European University Association [6], 12% of the polled institutions of higher education offer MOOCs, 42% plan to create such courses, and 39% of respondents do not plan to implement the MOOCs in the near future. Consequently, the introduction of e-learning in a variety of forms into the world of educational market is beyond doubt. Under this light the task of entering of Ukrainian educational institutions on this market is highly promising [9]. The actual picture of e-learning usage in Europe might be built on the basis of existing frontal statistics which is gathered by the European Statistical Committee [13]. The main drawbacks of that source are the following:

available statistics cover all European countries but the last year available is not the current one;
it is not specified (and could not be specified) to the higher education institutions sector—it covers all individuals (all adults who may obtain formal or informal education as well) or all students (all individuals who obtain formal education of ANY education level).

Regarding the drawbacks mentioned above it is possible to use the following data only in conjunction with the surveys of the European University Association and the EDUCAUSE, which were discussed previously.

One of the most self-explanatory indicators is the percentage of individuals doing any online course by the European countries (Fig. 1).

The actual doing of online courses by population of European countries shows the actual statistics from consumer point of view: the way the educational service actually is consumed. It is evident that students are involved in e-learning process much more

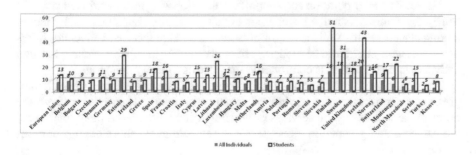

Fig. 1 Internet use: doing an online course (of any subject), in 2017 [13]

than overall population due to many peculiarities of educational service providers (not all the consumers may choose the way education course is delivered—in some cases the IT usage can't be avoided). The true front runners in delivering online courses are Scandinavian and Baltic countries with Iceland and United Kingdom.

The European Statistical Committee gathers on regular basis the statistics about the following types of e-learning activities [13]:

searching of online education opportunities, material, courses;
doing of an online course;
reading of online learning material;
communicating with instructors or students using educational websites/portals.

The aggregated values on all of those types of e-learning activities might be seen in Table 1, the last three (available) time periods were taken.

The data from Table 1 show the actual picture of usage of information technologies in education by European population in dynamics. On that basis the conclusion might be drawn that frontal usage of information technologies in education is status quo for European area—it is used by all strata of population, Western and Northern European countries are leaders in that process, Eastern European countries have a significant potential in it: their need of introduction information technologies in education is evident.

The quality education as one of global sustainability development goals have made many learning materials (free) open and available online. That opportunity is widely used by European population as a whole, and by students in particular (Fig. 2).

Serious activity of Western and Northern Europeans in online learning material search also might be explained with the fact that knowledge of prime global languages in those countries is high (most of population know English and/or German beside mother tongue). All that makes available online learning materials truly useful for consumers.

One more facet of actual information technologies usage in education is availability (or level of incorporation) of online social activities in education process: it might be peer education, online group activities, social media for education purposes, forums of all the kinds, etc. They reveal separate direction of pedagogic activities: social education (Fig. 3).

The level of incorporation and support of education activities mentioned above in European countries is much lower among adult population, but substantially higher among students. That statistics highlights the undeniable fact that University Information System in order to be modern, not outdated have to support social education activities in a way which is comfortable for students of every age group.

The considered statistic surveys make complete picture of information technologies usage (from the point of view of actual consumers) and provision (from the point of view of education services providers) in education. Even more, these data are gathered and grouped in global level so they help avoid shortcomings of cases surveys.

So, it should be noted that real pace of changes in global education sector (especially its private part) is much quicker than stereotypes. Thus, higher education

Table 1 Usage of Internet for any type of learning activity (on the basis of [13])

Country	Time					
	All individuals			Students		
	2015	2016	2017	2015	2016	2017
European Union—28 countries	17	18	19	54	55	59
Belgium	20	20	20	58	54	46
Bulgaria	10	9	8	55	52	52
Czech	8	11	13	33	53	59
Denmark	20	–	24	36	–	47
Germany	15	15	18	51	55	59
Estonia	34	39	31	83	92	87
Ireland	12	14	13	44	51	47
Greece	8	9	7	18	29	18
Spain	24	23	28	62	55	70
France	13	12	15	45	47	54
Croatia	12	17	10	36	51	50
Italy	15	15	16	57	56	58
Cyprus	7	10	10	23	36	35
Latvia	20	20	19	82	71	73
Lithuania	17	18	21	78	80	78
Luxembourg	33	34	37	54	67	80
Hungary	11	9	10	48	42	45
Malta	14	14	25	59	55	79
Netherlands	23	25	26	72	64	77
Austria	18	18	20	68	67	69
Poland	9	11	11	32	35	38
Portugal	20	23	24	60	66	70
Romania	20	14	17	56	41	51
Slovenia	16	15	18	53	49	49
Slovakia	13	12	13	46	41	45
Finland	29	30	31	89	89	92
Sweden	25	28	40	72	70	85
United Kingdom	27	29	30	75	70	72
Iceland	–	–	74	–	–	95
Norway	30	28	30	77	53	66
North Macedonia	16	13	9	45	36	24
Turkey	4	4	4	15	16	15

Note "–" means values are missed

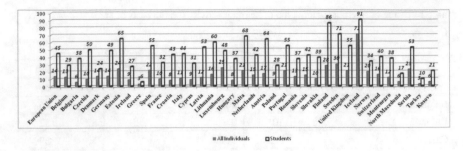

Fig. 2 Internet use: online learning material, in 2017 [13]

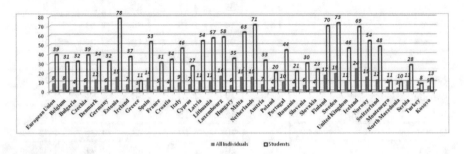

Fig. 3 Internet use: communicating with instructors or students using educational websites/portals, in 2017 [13]

institutions try to meet these changes, though in most cases higher education institutions' infrastructure and information systems unable to do that. Consequently, there is a pressing scientific task of redefining infocommunication needs of higher education institutions through their information flows in order to identify actual modern requirements to the higher education institutions' information systems and infocommunication infrastructure.

2 Background

According to significant research conducted by scientists such as: Kryvinska N., Gregus M., Molnar R., and others, the key to the development and implementation of a coherent information communication system is accurate analysis and detailed planning of communication processes. It should take into account the specifics of the industry, because universities provide a service in the form of education, which leads such activity to the services economy sector, the information flows in which should be thoroughly investigated and planned, and differ from the industrial sector [14]. Researchers insist that the university should promptly meet the internal technological and economic needs for the formation of sustainable and efficient infocommunication flows [15–19].

In continuation of the proposed research topic, Kirichenko L., Radivilova T., Zinkevich I., Dobrynin I, Maltseva N., Bulakh V., Tkachenko, A. and others note that when choosing a method of forecasting a company's business [20, 21], the service provides an important client-oriented approach, which strives to take into account the entire range of diverse interests of customers to the maximum, which will contribute to the formation of a wide network of information and communication flows directed from university to client, and accordingly adjusted by government influence and partners. At the same time, it is important to directly analyze information risks from qualitative to quantitative indicators [11], it allows authors to establish causal relationships of deviations obtained using factor analysis of information risks. In this respect, one should pay attention to the method of multifractal analysis of the influence degree of infocommunication networks on the financial dynamics [22–26].

As a result of research conducted by: Ageyev D., Al-Ansari A., Qasim N., Wehbe F, and Al-Dulaimi, A., determined that the rapid development of information technology leads to the need to create an information and communication system that provides the ability to transfer any type of information from anywhere in the world, at any time. Such infocommunication systems should differ from the previously used architecture and will not limit the number of services provided and the types of information transmitted [27]. Modern information and communication systems in their structure are multi-level. The processes occurring at different levels have a strong influence on each other. The structure of the system at one of the levels has the same strong influence on the characteristics of the other. Therefore, when solving problems of the effective functioning of such a system and parametric synthesis, the system should be treated as a single integral object [28–31].

In research of higher education institutions much attention is paid to information systems that support the learning process itself, among them: e-learning systems [32, 33], mobile learning systems, even MOOC's platforms (platforms for Massive Open Online Courses). But higher education institutions' infocommunication tasks are wider than support of distance and blended learning, they are not limited to education process. There are plenty of other infocommunication needs which should be supported with information systems and are currently overlooked in scientific literature (or scarcely discussed):

marketing communication [34];
business operations support [35];
governmental and funding authorities' communications;
decision support and providing analytics [27, 36].

Even more to say University Information System should satisfy the needs of various stakeholders that make such a system multilayered, with many subsystems. The latter fact may influence dramatically the stability of the system (i.e. see [37]). Thus, as far as University Information System is more than education one, it should incorporate elements (modules) that support specific needs of higher education institution (university) (i.e. in [38]).

3 Specific Higher Education Institution's Needs

Stakeholders in education, including government department heads, politicians, development agencies, community leaders, parents, students, and the general public, want to know what is happening in the education system. There is a growing need for universities to be accountable to supervision of both higher education authorities and local stakeholders (the trend of rising influence of non-governmental organization is widely discussed, i.e. in [39]). By actively sharing information, universities and other institutions in the education system can bring education closer to their local communities in order to improve understanding and participation, as well as mobilize their support for education for all.

Due to specific needs of higher education institutions, their info-communication system has to be build (or rebuild) in accordance with the variety of tasks resolved.

The list of specific higher education institutions infocommunication needs (but not limited to) is as follows:

take part in open educational projects (with free access to learning objects);
protect intellectual rights on produced educational and scientific objects;
support total social media presence;
support marketing activities;
avoid of duplicating of information (especially various learning objects);
provide formal reports to different stakeholders;
(peculiar) transparency of processes in order to attract extra funding from various authorities (especially nongovernmental);
support various learning activities: group work and peer education; virtual labs; gamification activities;
support augmented reality projects and virtual roaming;
support document workflow (especially, electronic workflow);
provide access to various libraries (including but not limited to: scientific library of university scholars, original textbooks and prominent fountainheads, etc.);
customize education services to peculiar user needs or societal demands [40];
meet standards of various authorities (i.e. governmental licenses, e-learning standards);
support partnership programs with enterprises and communities of practice.

Those new needs in most cases are supported by new information technologies, used in education sector.

4 Cutting-Edge Technologies in Education

The most expected, cutting-edge technological innovations in education and training are the following [2, 41]:

adaptation and personalization of learning through the use of artificial intelligence systems to identify learning progress and the most promising ways (for example, IBM's artificial intelligence platform Watson);

chatbots—softwareagents (works) for automatically generating chats that emulate human responses—they are designed to solve a learning problem as a scalable chat to meet the needs of any (by number) audience;

location-based Learning, which allows tracking the personnel behavior in real time and place (Real-Time Performance and Decision Support);

technology of psychometry for use in gamified learning;

technology of augmented reality and virtual reality. The augmented reality is the overlay of images, circuits, multimedia, 3D objects, location data, and other forms of digital content into real-world objects using a digital camera and sensors; most of the content of the augmented reality is "triggered" to certain objects (triggers): objects of recognition, printed markers, bar codes and geotags [2, p. 74];

cognitive learning technologies are meta-cognitive, since they enable the ability to change cognitive behavior (learning) through understanding and manipulating the learning process; in e-learning systems, cognitive learning technologies are "embedded" through appropriate technological tools (additional functions).

Let's consider some innovations in detail. Thus, games (or so-called "serious games") are becoming widespread in education and training; they allow learners to be trained in programmed situations and train relevant reactions and actions (see [42]). The reasons for their proliferation are high sensitivity and positive perception of the game submission of the learning material by the most diverse groups of learners, as well as the application of certain social and psychological technologies, such as psychometrics. Psychometry is a science that focuses on the statistical measurement of psychological states, and is therefore widely used to assess knowledge, abilities, skills, attitudes and personal qualities of staff [2].

The achievements of psychometrics are actively used to prepare and test learners for certification exams, as well as to train staff. Examples of such products and systems include Simcoach Method, Star Chart, Star Walk. The market for training and testing learners using the application of games (with the technology of psychometry) is considered as one of the most promising in the industry and is growing rapidly.

Technologically complex innovations that are fast-growing and becoming cheaper are technologies of augmented reality and virtual reality. Platforms of augmented and virtual reality are [2]: Junaio (developed by Metaio, now Apple), Cardboard viewer and Daydream virtual reality platform (developed by Google), HoloLens augmented reality platform (developed by Microsoft). They are complemented by software (hardware and software) complexes for automatic (or automated) spatial objects construction—Automatic Spatial Mapping tools: Producer Pro (developed by NGRAIN), Wikitude Studio (developed by Wikiddle), RealSense technology in Project Tango (developed by Google)—with Lenovo (Phab2 Pro), Eon Creator 7 and EON 9 (developed by Eon Reality), DinoTrek (developed by Geomedia, based on Google Cardboard). All of these platforms and technologies will reduce the cost and simplify the creation of educational media content. Augmented reality technologies are actively used to create tourist guides and museums guides, as well as to train learners to assemble technologically advanced objects (motors, turbines, aircraft and ships), for safety training. For example, the PTC company trains clients of the ERP and CAD platforms with the help of augmented reality. Blippar offers software that builds triggers of augmented reality on printed materials, which in turn allow to make "interactive" and "augmented" magazines, newspapers, and textbooks.

Cognitive technologies have a significant impact on the market of electronic training courses, since they fundamentally change the learning process. They are used in three main directions [2, p. 80]:

cognitive assessment—mainly consists of evaluating memory, temperament, ability to solve problems, spatial perception, as well as so-called "intentional" states of users. Indeed, the main application of the tool is in recruitment systems, but also very promising for testing entry level of students (in order to customize education services) and the results of education;
cognitive and intellectual counseling—aims at counseling and correcting person's behavior in real time and helping to improve cognitive abilities;
brain training and intellectual support—products (and educational services based on them) are widely used for medical diagnosis, as well as to stimulate sports achievements.

Special attention needs to be paid to innovations in the application of artificial intelligence technologies in university educational systems as far as they fundamentally change the learning process and some of its structural elements. At the moment, artificial intelligence is used in educational information systems to accurately identify gaps in knowledge, skills and abilities of learners, as well as the construction of a system of learning material in accordance with the personal needs of the learner. Moreover, such systems allow real-time changes in the content and in the dynamics of tasks provision.

Such systems are used in training soldiers, modeling complex production situations, as well as in language learning (for language courses). For the latter, artificial intelligence is used together with the latest developments in the field of automatic speech recognition, natural language processing, machine learning.

The practical application of programmable chatbots technology is currently being used by the provider of English language education services in China (English quiz), chatbots play a motivating role for students, encouraging them to study, and also assess current progress in the learning of the e-course. Similarly, the Indian firm PaGaLGuY uses chatbots in the education process.

Particular attention attracts systems that combine artificial intelligence and complementary technologies, creating powerful training facilities for jet aircraft (Lockheed Martin), ship building (Newport News Shipbuilding), air transport (Japan Airlines), and more.

5 Security Issues of University Information Systems

Introduction and massive usage of university information system presupposes dealing with security issues: some of them are common for any information system, some of them are specific (or even ad hoc) for higher education organizations due to particular characteristics of their business processes.

The safety aspects common to information systems are thoroughly researched [43–45], the methods and recommendations for their protection are proposed on the basis of scientific attainments [24, 46, 47]. So there is no use to dwell on the general security issues like disclosure, loss, leakage, service destruction.

The particular (to university education) process and influencing security factors are shown in Fig. 4 (taken from [33]).Those activities are common for many systems but for university information system they are critical because they hold all the main detected threats of University Information System: disclosure, integrity violation and denial of service [33].

If to pass-through the common activities in University Information System during single cycle of learning the generalized model might be obtained (Fig. 5).

Thus, University Information System have to incorporate general means of information security protection, as well as specific organizational and technical means of safe and confidential storage, usage and transfer of data [33].

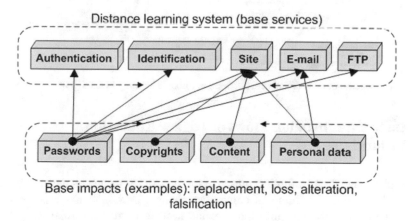

Fig. 4 Influence of security factors on the distance learning system [33]

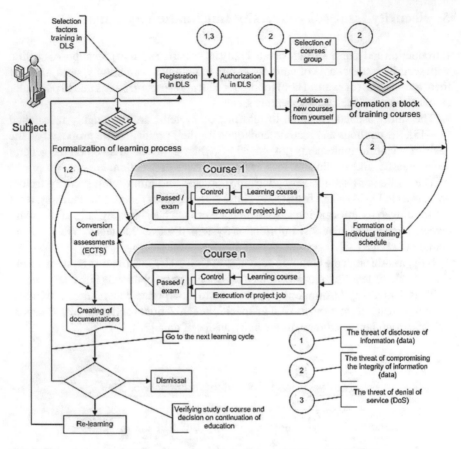

Fig. 5 The scheme of educational process cycle in distance learning system [33]

6 Infocommunication Flows of Higher Education Institutions

The higher education institution presence on the market means its development in extremely difficult conditions of continuous change. Certain characteristics of the marketing environment impose the constant need to correct information, process and analyze the data obtained in order to properly substantiate the optimal decisions related to the market [34].

There is no way an organization could work efficiently and effectively without information. Through the information flow managers can plan, direct, coordinate and motivate their staff. Similarly, managers receive feedback on the performance of their institutions from their clients through the fluent information flow.

Information flow refers to the physical movement of information from one team member to another or from one division to another. Any change of information is not considered as information flows. The information flow system is a collection of all physical information movements. Such a system makes it possible to carry out a process and implement a solution. The most common system of information flows is the sum of information flows that allows a commercial university to conduct financial and economic activity. Information flows ensure the normal operation of the educational institution. The purpose of working with information flows is to maximize the university's efficiency.

Managers can control the information flows as follows:

changing the direction of flow;
limiting the transmission rate to the corresponding reception rate;
limiting the stream volume to the value of an individual node capacity or the path section.

The main condition for the management process is the processing of information circulating in system. The information flow can be ahead of the service providing, to follow simultaneously with it or after it. In the case, the information flow can be directed both in one direction with the educational service and in the opposite direction. The path along which the flow of information, in general, may not coincide with the route of the service providing flow.

The information flow is characterized by the following indicators: the source of occurrence; flow direction; transmission and reception speeds; flow intensity, etc. The moving information flow in the opposite direction contains, as a rule, information about the order (for some educational product). Simultaneously with the educational flow, there is information in the forward direction about the quantitative and qualitative parameters of the educational process flow.

The formation of information systems is impossible without the study of flows in the context of certain indicators. The documents used in the management process may include one or more indicators with a mandatory certification (signature or seal) of the person responsible for the information contained in the documents. Since the acquisition of source data is the field of human activity, most of the documents are created at the stage of data collection and registration, although a large proportion of the documents come into the system from external (superior and other) organizations.

Improvement of the information transfer process, in the case, it is a question of managing communications in each separate business process. An organization needs not only to collect, store and transfer (process) information, it needs to know who is responsible for the quality of the information transfer. The task of quality within the framework of any business process is designed to be solved by professionals (who have got specialized education in this field). For them, information is data that reduces the uncertainty of the consumer's knowledge about certain objects or processes (for example, business), and as a result, all their efforts are to implement informational aspects: creating and maintaining corporate information funds, preparing analytical materials, creating advertising materials, products in their own direction. At the same time, very often their actions are not coordinated, for example, an internal corporate communications specialist may simply not be aware of the activities carried out by the marketing communications specialist, and the PR specialist may not know about the programs conducted by the HR specialist. These problems often arise from the fact that each of the above-mentioned specialists implements their functions within the unit in which he works, and the general corporate strategy "in the field of information and communications" is absent or is known only to the "first person". The work of a specialist in the field of information and communications is monitored by the head, who manages a specific business process and whose goal is to achieve key performance indicators. Since the activity of any organization is aimed at optimizing the resources use and therefore the management of the company is faced with the issue of information management.

Information management is a process of management, not only people possessing information but also actions allowing to select the business processes chains under which organizational structures and IT support are adjusted. In this case, information management is one of the few processes in an educational organization that exists within the framework of rigidly defined constraints, such as:

- An information need arrives at the input of the process, and the output is a generated information service that is provided to the customer (student) and the end user.
- The process proceeds independently of the organizational structure and functional business tasks/operations.
- The information management process can also proceed in the opposite direction: IT infrastructure analysis is taking place from the generated service and a new need is revealed that has not been considered before or has been unclaimed.

This goal is achieved in the process of managing information (Information Management Process) which begins (at the entrance) with the identification or receipt of the customer's (student's or other stakeholders) information needs, and at the output, a list of the means and data necessary to achieve the goals is formed. The task, in this case, is to identify this process, its regulation, reengineering, as well as the creation and support of basic information services. Specialists (managers) in the management of information flows are responsible for managing this process, and all this is implemented through the introduction of a University Information System (UIS), while it is an integral part of the overall management system.

For this, IT specialists introduce systems: for managing relationships: with customers—students (CRM) and partners (BTB), with personnel (HRM), government (BTG) for organizing information flows they introduce electronic workflow, etc. (Fig. 6).

The role of UIS in decision making in a University Information System (UIS) allows an organization to combine previously separated techniques that were subprocesses in other business processes and form an independent business process based on them that has not only input and output but also makes it possible to evaluate profits from its functioning. One often declare that information is money, but constant control over the use of this resource is possible only with the implementation of an information management system.

In practice, often the lack of control over information flows is not due to the management's unwillingness to control this process, but with the lack of information about their management practices. University Information System (UIS) obliges the

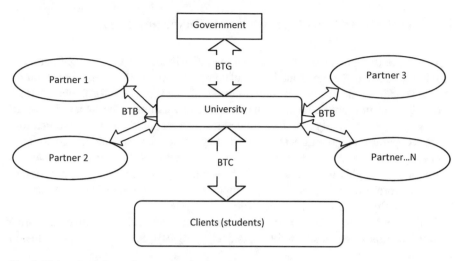

Fig. 6 University information communication system

educational organization's management to manage information by planning their needs, providing the communication channel for information transmission, monitoring its profitability and efficiency of its consumption. Plans in it are formed on the basis of analysis of needs, and the process of information structure managing implies managing not only IT support tools (technical, software and organizational subsystems), but also through other channels of information distribution. At the same time, at the output of the management process information flow, received information about a complex of streamlined subsystems, in which the control is carried out by comparing it with the prescribed standards. In this case, information flows not only work for the customer (student), providing the satisfaction of the information need but also allow responsible persons to control the process, i.e. in the process of customer relationship management, the process of improving the system becomes an important subprocess. The comments to the system building are formed in the UIS, questions for refinement and technical support are visible, as well as many other activities related to information flows.

This principle of the information management system control can be implemented in almost any commercial educational infrastructure since the general output of the information management process is an educational service that is created and operates for the customer. Moreover, not only the owners of the process—specialists in the field of information management but all employees of the educational organization who have to make management decisions are engaged in daily work in this system. Thus, when introducing UIS within any organization, explicitly or implicitly, there is a process of changing not only the management system but also the corporate culture by introducing information management, that brings higher quality and efficiency of educational process.

7 Future Prospects of University Information Systems

According to the specific infocommunication flows in universities, it is obligatory to provide basis for all the mentioned types of communications. But even more attention should be paid to the problem of integration of all the types of information systems currently used in higher education institutions (universities). The integration of all main groups of university activities (education process; marketing communication; business operations; governmental and funding authorities' communications; decision support and business analytics) under umbrella of solid information system is a complex and highly difficult task. It implies not only integration of many different existing information systems, but more important thing is the set of data that need to be used simultaneously in different information systems.

The issues of integration in education information systems and university information systems where discussed (respectively [48, 49]), but no unified conclusion was drawn.

In order to resolve the problem, it is possible to provide all systems integration or make a refactoring. Let's consider those options from the point of author's experience of university information system management.

The most often used method is system integration, that implies remaking (improvements) of currently used information systems. That decision is supported by widely used idea of supporting current infrastructure as it is, and management point of view that it is the most natural way of university information systems development. But the shortcuts of this decision is also evident (as well as widely faced)—at some point the system cannot support all the needs of higher education institution, so most likely university will reject some innovative (and highly perspective) information technologies due to those pitfalls. In some cases, university can choose "a middle" alternative—aggregate all the data in one data-warehouse. It gives an opportunity to partially resolve the task of data analytics and support business decisions, but doesn't resolve the problem entirely.

The other way of development of university information system (in order to meet modern needs) is to make a system refactoring, i.e. using microservice architecture. Such decision is more expensive in short-term period but gives an important strategic benefit—it gives an opportunity to incorporate modern information technologies in university information infrastructure in order to provide highly valuable education service.

8 Conclusion

Thus, the scientific novelty lies in the fact that the University Information System (UIS) proposed in this study involves the key elements system integration (a university, clients—students, partners and government) into a united space with converging information and communication flows. This will make it possible, on the one hand, to significantly simplify and optimize the sides' cooperation—due to the fragmentary inclusion of certain participants into the general information flow during their presence in the business process. On the other hand, it will significantly increase the system functioning efficiency, accordingly, reducing its implementation and provision costs.

Many higher education institutions need to improve their information systems and infocommunication infrastructure in order to meet modern specific needs of education sector. Therefore, university information system integration challenge looms on the horizon of the most higher education institutions. The conclusions of this work may provide a solid ground for such a decision.

References

1. Kavun S, Mykhalchuk I, Kalashnykova N, Zyma A (2012) A method of internet-analysis by the tools of graph theory. In: Watada J, Phillips-Wren G, Jain LC, Howlett RJ (eds) Advances in intelligent decision technologies. Smart innovation, systems and technologies, vol 15, Part 1. Springer, Heidelber, Germany, pp 35–44. https://doi.org/10.1007/978-3-642-29977-3_4
2. Adkins SS (2016) The 2016–2021 worldwide self-paced eLearning market: the global eLearning market is in steep decline. http://www.ambientinsight.com. Last accessed 14 Apr 2019
3. E-Learning Market Trends & Forecast 2017–2021 Report (2016) A report by Docebo. Docebo S.p.A, London
4. Report and Forecast to 2017–2022 (2017) A report by Orbis Research. Dallas, Texas, June 15 2017. http://www.orbisrescarch.com/contacts/request-sample/226355. Last accessed 14 Apr 2019
5. Sucre F (2016) Role of E-learning in higher education in Latin America. Inter-American Dialogue, 12 Apr 2016. http://www.thedialogue.org/2016/04/role-of-e-learning-in-higher-education-in-latin-america. Accessed 14 Apr 2019 (2016)
6. Gaebel M, Kupriyanova V, Morais R, Colucci E (2014) E-Learning in European Higher Education Institutions. Results of a mapping survey conducted October–December 2013. European University Association. EUA, Brussels
7. Sursock A (2015) Trends 2015: learning and teaching in European Universities. European University Association. EUA, Brussels
8. Bichel J (2013) The state of E-Learning in higher education: an eye toward growth and increased access (research report). CO: EDUCAUSE Center for Analysis and Research. EDUCAUSE, Louisville
9. Zavgorodnia O, Mikheev I, Zyma O (2018) Identifying European E-Learner profile by means of data mining. In: IEEE second international conference data stream mining & processing 2018, 21–25 Aug 2018, Lviv, pp 202–206
10. Kavun SV, Brumnik R, Kalashnikov VV, Kalashnykova N (2017) Security aspects in critical infrastructures. In: Gorbenko ID, Kuznetsov AA (eds) ISCI'2017: information security in critical infrastructures. Collective monograph. LAP Lambert Academic Publishing, OmniScriptum GmbH & Co. KG, Germany, pp 7–44
11. Panchenko V, Zamula A, Kavun S, Mikheev I (2018) Intelligent management of the enterprise personnel security system. In: 2018 IEEE 9th international conference on Dependable Systems, Services and Technologies (DESSERT), Kiev, pp 469–474. https://doi.org/10.1109/dessert. 2018.8409179. (INSPEC Accession Number: 17917430). Available from https://ieeexplore. ieee.org/document/8409179/

12. NVAO MOOCs and Online HE (2014) A survey conducted June 2014, Accreditation Organization of the Netherlands and Flanders (NVAO). NVA, The Hague
13. Eurostat database (2019) http://ec.europa.eu/eurostat/data/database
14. Tatar M, Sergienko O, Kavun S, Guryanova L (2017) Complex of management models of the enterprise competitiveness for steel industry in the currency instable environment. Econ Stud (5)
15. Kryvinska N (2012) Building consistent formal specification for the service enterprise agility foundation. J Serv Sci Res 4(2):235–269 (The Society of Service Science, Springer)
16. Kaczor S, Kryvinska N (2013) It is all about services—fundamentals, drivers, and business models. J Serv Sci Res 5(2):125–154 (The Society of Service Science, Springer)
17. Gregus M, Kryvinska N (2015) Service orientation of enterprises—aspects, dimensions, technologies. Comenius University in Bratislava
18. Kryvinska N, Gregus M (2014) SOA and its business value in requirements, features, practices and methodologies. Comenius University in Bratislava
19. Molnar E, Molnar R, Kryvinska N, Gregus M (2014) Web intelligence in practice. J Serv Sci Res 6(1):149–172 (The Society of Service Science, Springer)
20. Kavun S, Zhosan G (2014) Calculation of the generalizing indicator of productivity of the enterprises activity based on the matrix-rank approach. J Finance Econ 2(6):202–209. https://doi.org/10.12691/jfe-2-6-1
21. Kavun S, Zhosan G (2016) Influence determination of social responsibility to the productivity enterprise activity level. Economies 4(3):14. https://doi.org/10.3390/economies4030014
22. Kirichenko L, Radivilova T, Zinkevich I (2017) Forecasting weakly correlated time series in tasks of electronic commerce. In: Proceedings of 2017 12th international scientific and technical conference on Computer Sciences and Information Technologies (CSIT), Lviv, pp 309–312. https://doi.org/10.1109/stc-csit.2017.8098793
23. Kirichenko L, Radivilova T, Zinkevich I (2018) Comparative analysis of conversion series forecasting in e-commerce tasks. In: Shakhovska N, Stepashko V (eds) Advances in intelligent systems and computing II. CSIT 2017. Advances in intelligent systems and computing, vol 689. Springer, Cham, pp 230–242. https://doi.org/10.1007/978-3-319-70581-1_16
24. Dobrynin I, Radivilova T, Maltseva N, Ageyev D (2018) Use of approaches to the methodology of factor analysis of information risks for the quantitative assessment of information risks based on the formation of cause-and-effect links. In: Proceedings of the 2018 international scientific-Practical Conference Problems of Infocommunications. Science and Technology (PIC S&T), Kharkiv, Ukraine, pp 229–232. https://doi.org/10.1109/infocommst.2018.8632022
25. Kirichenko L, Bulakh V, Radivilova T (2017) Fractal time series analysis of social network activities. In: Proceedings of the 2017 4th International Scientific-Practical Conference Problems of Infocommunications. Science and Technology (PIC S&T), Kharkov, Ukraine, pp 456–459. https://doi.org/10.1109/infocommst.2017.8246438
26. Kirichenko L, Radivilova T, Tkachenko A (2019) Comparative analysis of noisy time series clustering. In: Proceedings of the 3rd international conference on Computational Linguistics and Intelligent Systems (COLINS-2019), vol I. Kharkiv, Ukraine, 18–19 Apr 2019, pp 184–196
27. Zamula A, Kavun S (2017) Complex systems modeling with intelligent control elements. Int J Model Simul Sci Comput 8(1). https://doi.org/10.1142/s179396231750009x
28. Ageyev D (2010) NGN network planning according to criterion of provider's maximum profit. In: Proceeding of the IEEE international conference on Modern Problems of Radio Engineering, Telecommunications and Computer Science (TCSET-2010), pp 256–256
29. Ageyev D, Al-Ansari A, Qasim N (2015) Multi-period LTE RAN and services planning for operator profit maximization. In: Proceeding of the IEEE the experience of designing and application of CAD systems in microelectronics, pp 25–27. https://doi.org/10.1109/cadsm.2015.7230786
30. Ageyev D, Wehbe F (2013) Parametric synthesis of enterprise infocommunication systems using a multi-layer graph model. In: Proceeding of the IEEE 23rd international crimean conference microwave & telecommunication technology, pp 507–508

31. Al-Dulaimi A, Al-Dulaimi M, Ageyev D (2016) Realization of resource blocks allocation in LTE downlink in the form of nonlinear optimization. In: Proceeding of the IEEE 13th international conference on Modern Problems of Radio Engineering, Telecommunications and Computer Science (TCSET), pp 646–648. https://doi.org/10.1109/tcset.2016.7452140

32. Pushkar O, Zavgorodnia O (2018) Mind map usage for structuring information space of learning disciplines in E-learning. Syst Inf Process 2(153):108–116. https://doi.org/10.30748/soi.2018. 153.14

33. Kavun S, Daradkeh Y, Zyma A (2012) Safety aspects in the distance learning systems. Creative Educ 3:84–91. https://doi.org/10.4236/ce.2012.31014

34. Casap L (2018) The informational support in educational marketing decision-making process in the Republic of Moldova. Mark Inf Decis J Sciendo 1(2):5–11. https://doi.org/10.2478/midj-2018-0006

35. David G, Ribeiro LM (1999) Getting management support from a University Information System. In: Proceedings of the European cooperation in higher education information systems, EUNIS99, Espoo

36. Rakic K, Bandic GM, Majstorovic V (2014) Development of the intelligent system for the use of University Information System. Procedia Eng 69:402–409

37. Goykhman M, Kavun S (2014) Evaluation method of banking system stability based on the volume of subsystems. J Finance Econ 2(4):118–124. https://doi.org/10.12691/jfe-2-4-3

38. Bassil YA (2012) Data warehouse design for a typical university information system. J Comput Sci Res (JCSCR) 1(6):12–17

39. Holovatyi M (2015) The state and society: the conceptual foundations and social interaction in the context of formation and functioning of states. Econ Ann XXI(9–10):4–8

40. Holovatyi M (2014) Multiculturalism as a means of nations and countries interethnic unity achieving. Econ Ann XXI(11–12):15–18

41. Zavgorodnia O (2019) Analysis of trends in E-Learning. In: International Forum on Information Systems and Technology (INFOS-2019), Kharkiv, pp 26–28

42. Zavgorodnia O (2016) Economic security aspects of pervasive gamification. In: III international conference on Information and economic security (INFECO-2016): Kharkiv, pp 125–127, 28–30 Apr 2016

43. Kavun S, Brumnik R (2013) Management of corporate security: new approaches and future challenges. In: Galeta D, Vrsec M (eds) Cyber security challenges for critical infrastructure protection. Institute for Corporate Security Studies, Ljubljana, pp 141–151

44. Kavun S, Sorbat I, Kalashnikov V (2012) Enterprise insider detection as an integer programming problem. In: Watada J, Phillips-Wren G, Jain LC, Howlett RJ (eds) Advances in intelligent decision technologies. Smart innovation, systems and technologies, vol 12, Springer, Heidelber, pp 820–829. https://doi.org/10.1007/978-3-642-29920-9_29

45. Kavun SV, Kalashnikov VV, Kalashnykova N, Cherevko O (2015) Evaluation of financial losses suffered by enterprises due to information system accidents. In: Ohnishi K (ed) Firms' strategic decisions: theoretical and empirical findings, vol 2, Chap. 10. Bentham Science Publishers, Sharjah, pp 239–261

46. Kavun S (2015) Conceptual fundamentals of a theory of mathematical interpretation. Int J Comput Sci Math 6(2):107–121. https://doi.org/10.1504/IJCSM.2015.069459

47. Kavun S (2016) Indicative-geometric method for estimation of any business entity. Int J Data Anal Tech Strat 8(2):87–107. http://dx.doi.org/10.1504/IJDATS.2016.077486

48. Jakimoski K (2016) Challenges of interoperability and integration in education information systems. Int J Database Theory Appl 9(2):33–46

49. Ise OA (2014) Towards a Unified University information system: bridging the gap of data interoperability. Am J Softw Eng 2.2:26–32. Last accessed 14 Apr 2019

Advanced Methods and Means
of Authentication of Devices
for Processing Business Information

Elena Nyemkova⬤, **Vyacheslav Chaplyha**⬤, **Jan Ministr**⬤
and **Volodymyr Chaplyha**⬤

Abstract The article is devoted to the analysis of the current state of methods and means of authentication of electronic devices for tasks of processing business information. Transmission, transformation and storage of business information are carried out by a wide spectrum and increasing number of devices both in global networks and outside of them. This requires adequate mutual authentication between devices to enhance cybersecurity, as well as reliable authentication of a variety of electronic devices for digital media expertise, detection of counterfeit products, etc. The general approach to the authentication of electronic devices based on their individual differences allows solving various applications of business informatics. The classification of such tasks, as well as the analysis of methods and means of authentication of electronic devices for solving various classes of tasks showed the following. Forensic tasks are solved by methods of comparing two samples from two random processes based on Kolmogorov-Smirnov's agreement with the use of specialized software such as "Fractal". For geolocation tasks of an electronic device, it is expedient to use individual template for an error vector in determining the deviations in the signal constellation. The tasks of detecting counterfeit notebooks and mobile phones are solved by methods of comparing the spectra of their own electromagnetic radiation of these devices with the standard. Remote automated real-time authentication of various Supervisory Control and Data Acquisition system sensors can be solved by passive analysis of the sensor's own noise based on its spectrum.

Keywords Authentication · Electronic device · Business information ·
Cybersecurity · Template · Forensic · Geolocation · Counterfeit products

E. Nyemkova (✉) · V. Chaplyha
National University Lviv Polytechnic, Lviv, Ukraine
e-mail: cyberlbi12@gmail.com

J. Ministr
VŠB—Technical University of Ostrava, Ostrava, Czech Republic

V. Chaplyha
Banking University Lviv Institute, Lviv, Ukraine

© Springer Nature Switzerland AG 2020
D. Ageyev et al. (eds.), *Data-Centric Business and Applications*,
Lecture Notes on Data Engineering and Communications Technologies 42,
https://doi.org/10.1007/978-3-030-35649-1_10

1 General Approach to Authentication of Electronic Devices

The development of the digital economy in individual countries, further integration of the digital markets of the Eastern Partnership countries into the single European space are related to the processing of large volumes of diverse business information and the corresponding development of the IT infrastructure [1–4]. At the same time, one of the main problems is to increase confidence and security in the digital economy, electronic government (e-government), etc. [5–8].

Electronic devices are used to transmit, transform and store information, they are widely used in modern life to create new values and extend the comfort zone [9–14]. Basically, these are electronic devices that interact and exchange information on the network; their diversity and quantity is constantly increasing. The We Are Social analyst and Hootsuite SMM Platform in the 2018 joint report quoted data from the global digital market [15], according to which the number of connected IPs in 2018 exceeded 4 billion. This significantly increases the risk of vulnerability to cyberattacks and leads to the need for a reliable mutual authentication between devices. The current device authentication system based on IP and MAC addresses is not too reliable. For example, the MAC address specified at the software level by the driver has precedence over the hardware level. This makes it possible to substitute the MAC address, which is the hardware characteristic of the network card. The volume of cybercrime grows from year to year, both quantitatively and qualitatively.

There is also a class of tasks where you need to authenticate an electronic device that is not necessarily connected to the information network. Basically, it is forensic tasks, or the detection of counterfeit or counterfeit products.

The variety of electronic devices provides a wide range of opportunities for using their individual differences for authentication purposes. The basic requirements for authentication solutions are practicality (automatic, real-time, remoteness), economic justification and reliability.

Recently, there has been an increase in the interest of researchers in the systems of authentication of electronic devices based on their individual differences. The production of electronic devices ensures their functioning within the limits of fault-tolerance and does not have the purpose of physical identity. Minor allowable deviations in the technological process of manufacturing integrated circuits lead to uneven distribution of acceptor or donor impurities, which determines the specificity of internal electrical noise when working with chips. Also, in the process of assembling electronic devices, the equipment with certain permissible electrical parameters is used, which also affects the output electrical signals or the specifics of internal electrical noise. Thus, each electronic device has physical features that can be used to authenticate it.

For the tasks of authenticating electronic devices it is accepted to use concepts, terminology and algorithms, similar to human biometrics [16–19]. First of all, this applies to the general approach to the device authentication process by the template. Template T (signature) is the display of measurements of the physical quantities under which the device is authenticated to a set of N numbers. The template $(\{T_i\}, i = \overline{1, N})$ must be individual for each device and stable for the same type measurements. In practice, the sign on which authentication is carried out may change insignificantly depending on unforeseen external and internal conditions, which leads to changes in the current pattern—for each measurement procedure l, they receive their set $\{T_{i,l}\}$. Therefore, as a rule, the process of averaging a template in a series of L measurements is carried out:

$$T_i = \overline{T_{i,l}} = \frac{1}{L} \sum_{l=1}^{L} T_{i,l} \tag{1}$$

This leads to the fact that the pattern of the device received in subsequent measurements may not coincide with the averaged template. To pass the authentication, the device introduces the threshold D—the maximum permissible deviation between the averaged and current patterns of one device.

By the rejection of two patterns, or conversely, similarities, one of four types of coefficients [20] can be used for different practical tasks: correlation coefficients, distance measures, associativity coefficients, and probability coefficients of similarity. Of the listed coefficients for the authentication tasks most often use forms of a special class of metric distance functions, known as metrics of Minkowski

$$d_{Tm1,Tm} = \left(\sum_{i=1}^{N} |T_{m1,i} - T_{m,i}|^r \right)^{1/r} \tag{2}$$

For the Euclidean metric $r = 2$, for the Heming distance $r = 1$.

Also, the Mahalanobis distance is used, which takes into account the correlation between the variables $\{T_{i,l}\}$. If correlation is absent, then the distance of Mahalanobis is equivalent to the distance of Euclidean.

Let $\{m\} = M$ is the various devices that need to be authenticated. For each m device, measurements are performed at the beginning of the authentication procedure and a template Tm is created which is then stored in the database and may be provided for comparison on demand. Thus, the database contains a set of templates $\{Tm\}$. If you need to authenticate device $m1$, you need to again measure physical signals and get the template $T1m1$ for comparison. The following authentication procedure depends on what needs to be done: to identify the device—to identify one of many (One-to-Many [21]), to verify it—to confirm that it is the same device (One-to-One [22]), or to check the presence of a template of this device in the database—to conduct Open Identification (Open-Set Identification [23]). A typical identification system performs the recognition of one electronic device $m1$ from many M based on the minimum distance between template pairs, one of which is the $T1m1$ pattern, and the second is taken from the $\{Tm\}$ set:

$$d_{T1m1,Tm} \to \min, m = \overline{1, M} \tag{3}$$

In this case, $m1$ is authenticated as m device for which the top condition is met.

The verification threshold is used to verify the device. If the distance between the $T1m1$ and Tm patterns is not greater than D, then the $m1$ device is authenticated as m:

$$d_{T1m1,Tm} \leq D \tag{4}$$

If you want to first check that there is $m1$ device template in the $\{Tm\}$ database and, secondly, to authenticate the device in the event of a successful validation, there are M comparisons:

$$d_{T1m1,Tm} \leq D, m = \overline{1, M} \tag{5}$$

If there is template Tm, for which the last condition is found, then the $m1$ device is authenticated by the system as the m device.

The individuality of each electronic device is unmistakable today and is confirmed experimentally. When authenticating electronic devices, there are two problems. The first is the availability of authentication means, that is, the availability of hardware for detecting and measuring individual differences that are realized in the form of physical signals. The second is the availability of methods that, based on measurements, implement algorithms for creating templates, their comparison and decision making regarding the authentication of the electronic device.

The combination of methods and means should produce a statistically repeatable result that can be characterized by the following four options: true positive identification, true negative identification, false positive identification (FRR, False Recognition Rate) and false negative identification (FAR, False Acceptance Rate is a mistake of the second kind). To characterize biometric systems it is customary to give the dependence of FRR and FAR values on the recognition threshold.

The accuracy of any authentication system should be characterized by two variables: the magnitude of the error of the first kind at a given error value of the second kind. The values of FRR and FAR also depend on the size of the sample—the number of objects of authentication. The higher the sample, the more reliable the result can be obtained with respect to recognition quality. Therefore, knowledge of only the threshold value is not enough, it is necessary to specify the number of checked objects.

Note that authentication of electronic devices using individual differences in their physical signals in real time requires obtaining and studying their integral characteristics. For this, the authors in [24–26] suggested algorithms for the deterministic non-equidistant digital filtering in real time of signals in the time domain based on "Codes with Minimal Aperiodicity" allow combining the positive properties of both periodic and stochastic processes. The minimum average sampling rate, the possibility of parallelizing computations in correlation-spectral analysis, and the tight binding of samples to the time axis are significant advantages of such a deterministic process. The latter is necessary in those cases when it is required to reconstruct the waveform from the set of samples. These algorithms allow to easily parallelizing the structure of the computing device, they are inherent in the simplicity of address calculation devices and, as a consequence, the simplicity of the control device.

2 Classes for Authentication of Electronic Devices

The variety of tasks for which the authentication of electronic devices is carried out is explained by a rather large list of methods and means for solving them. It is possible to share all methods and means according to the classes of tasks that must be solved.

1. A class of forensic or digital examination tasks, which is part of the proof of the authenticity of media files. The possibility of using such files as evidence has to be substantiated by the use of specific media recorders, which can be represented by digital microphones, digital and analog video cameras, mobile phones, WEB cameras, car video recorders, and computers more rarely. A very important requirement is the lack of compilation of media files, including the lack of digital post-processing to provide a record of certain properties or the introduction of additional information. To this item of classification can be attributed tasks of proof of copyright to the authentic media content.

2. Geolocation of the electronic device. The need to determine the exact geolocation is due to security requirements when authenticating the user—the owner of the electronic device in corporate networks to solve various tasks, for example, accessing databases of the university library's electronic library, conducting financial transactions through on-line banking services or directly using post-terminals or ATMs, and so on.

3. Detection of counterfeit notebooks (mobile phones) or their counterfeits is a class of tasks related to the financial and image interests of leading manufacturers of modern high-quality electronic equipment, and may also relate to the protection of consumer rights.

4. Remote automated real-time authentication of various SCADA systems (Supervisory Control and Data Acquisition), or sensors located in places with a high risk for a person; systems of automatic control of critical infrastructures—electric grids for domestic consumption, electric transport, and more. Acquiring a rapid development of the paradigm of intelligent human environment—there are smart houses and cities. The average number of names of different sensors only in a reasonable house can exceed dozens of units. On the basis of sensor displays, the control of climate control systems is maintained, systems are maintained in normal mode or an alarm signal is received.

5. Remote automated real-time authentication of personal computers or laptops in corporate networks. Solving the tasks of this category will enable:

   - Increasing the security of information exchange in corporate networks with role access to resources,
   - Continuous inventory of computers in wireless networks with restricted access as a means of information security audit,

– Increasing the security level of the Enterprise Resource Planning System (ERP) systems, the reliability of the DLP (Data Leak Prevention System) and other information systems,
– Increased security of financial transactions.

6. Authentication of integrated circuits, which may be integral parts of the control unit for vehicles, household appliances, etc. Authentication methods are based on the use of either physically non-dominant functions, or the characteristic radiation of an electromagnetic field.

The methods and means of authenticating electronic devices to provide different classes of tasks can vary greatly. The basis of the diversity of methods and means of authentication are both the physical causes of the occurrence of individual differences in electronic devices, as well as the methodology for their recognition. Physical reasons for the existence of individual differences in devices consist in the non-necessity of their absolutely exact copies at the physical level—for the standard operation of serial devices; a guarantee of work within the fault-tolerance is required. The methodology for identifying individual characteristics is related to the requirements for the speed of the recognition, the possibility of remote conduct, full automation or with the participation of a human expert, and mainly consists in identifying the statistical characteristics of the noise of registered signals, not related to useful signals, that is, such noises, made by the electronic device itself or the environment. Detection of features is carried out using spectral correlation analysis, as Fourier transforms or wavelets are most often used as orthogonal transformations to obtain periodogram (for brevity called spectra). Also known are other methods of identifying some non-noise devices.

Let's consider in more detail the classes of tasks of authentication of electronic devices from the point of view of the applied methods and means.

3 Authentication of Media Equipment for Digital Examination Tasks

Methods and means of authentication of media equipment for the tasks of digital examination are most fully presented for familiarization in comparison with other classes of authentication tasks [27–34]. The paper [35] presents modern systems designed for phonoscope examination, which have already been implemented or implemented in expert institutions of Ukraine. The basis for authentication is the physical differences of media equipment, which manifests itself as an additive noise of a digital image or audio recording overlaid on a useful signal when recording audio and video recordings or photographs. The technology for checking the identity of the audio equipment is based on the assumption that the noise is a random stationary process, so it is necessary to compare the characteristics of two random stationary processes. The most complete random process can be characterized by the distribution function. A comparison of two choices from two random processes for a situation where the process of distribution of processes is unknown can be carried out according to Kolmogorov-Smirnov's consent criterion. The identification algorithm consists of the following steps [34]:

(1) To isolate the noise from the pause of the speech signal of the phonograms.
(2) Selection of multifractal structures by the method of maxima of wavelet transformations (a large number of similar fractal structures of different levels guarantees the statistical ability of evaluations). Determination of probability distribution of degree of proximity of fractal structures fixed on different equipment of sound recording.
(3) Application of non-parametric criterion of agreement Kolmogorov-Smirnov to determine the membership of the two distributions received in the previous paragraph to one law.
(4) On the basis of the obtained results, the statistical hypothesis about the membership of the received functions of distribution to one law is accepted or omitted.

The automated system of identification of the equipment of analogue and digital sound recording "Fractal" developed on the basis of this algorithm provides high reliability of the results of examination. This system allows forensic identification of sound recording equipment, to establish the originality of phonograms and to detect traces of digital processing in them.

In the study of other authors, the identification of digital microphones was carried out by means of correlation of a pair of averaged over many audio recordings of noise spectra [36].

Many audio recordings are recorded by devices that are connected to a power supply. Normally, the rated frequency of the power supply is either 50 or 60 Hz. It is established that when connecting to the network of various capacities of consumers, there are slight deviations of the power frequency from the nominal value up to 0.1% [37]. These random vibrations are recorded in audio files and can be extracted for comparison with the known reference "trajectory" of the power supply changes [38–41]. For the last decade, criminology has used the value of random fluctuations in the frequency of the power supply (ENF), as the natural signature in many audio recordings [42–44]. ENF-based solutions can be used in cases of confirmation or rejection of compilation of audio files, or proof of the recording of a specific audio file in a more or less definite geographic region at a specific time [45], attracting both academic researchers and law enforcement agencies. For a successful application of the method, it is necessary to create a reliable base of reference data ENF with anchoring to the geographical position [46]. Although the idea of ENF does not directly relate to the authentication of specific audio devices, it can be said that it allows you to confirm the recording time for a particular geographic location of the recording device. There are several known algorithms that allow the proof of audio files based on the idea of ENF. One of these algorithms consists of two blocks [47]:

(1) Extraction of the sequence of random values of the ENF fluctuations from the audio file through the use of narrow-band digital filters for three harmonics of 50/100/150 Hz or 60/120/180 Hz.

(2) A procedure for comparing the sequence of random values of the ENF fluctuations with the reference sequence in accordance with the timestamps and geographic locations using the correlation function.

This algorithm was applied to an experiment in which the voltage from the power supply through the divider was fed to the linear input of one computer and simultaneously recorded the audio file on the second computer. The experiment received a 99% match between the ENF signal extracted from the test audio record and the ENF reference signal.

The use of the ENF method is complicated by the fact that the spectral range of the subject under study enters the own noise of the audio recorder and the spectral components of the sound. There is also a need to maintain ENF reference databases for different geographic regions.

As a rule, there is enough time for conducting a digital examination of audio files, so the requirement of working in real time is not critical.

4   Authentication of Mobile Phones for Geolocation Tasks

In the task class for determining geolocation for mobile phones, two tasks can be considered:

(1) The determining the geolocation of a mobile phone for solving digital criminology issues,

(2) The receiving information about the location of the user in his attempts to login to the Internet service system (relative proximity of the location of the mobile phone as the factor of authentication and a specific computer).

The 802.11 standard [48] defines the method of encoding information for its transmission via mobile communication. The output signal in the mobile phone is initially formed in digital form and passes through a digital-to-analog converter. Further, using the GMSK (Gaussian Minimum Shift Keying) technology, the signal passes through a band pass filter, mixer and amplifier. The symbol as a sequence of bits of the source information is converted into a modulated radio frequency signal, which is conveniently described using a signal constellation diagram. The radio frequency transmitter must perform decoding and character recognition bit sequences recognition. Typically, the complex vector of the input signal has a small distortion, which is expressed in its deviation—the error from the initial desired position on the signal constellation diagram for this combination of bits. Correction to the nearest position on the signal constellation diagram based on the Euclidean metric distance is corrected to correct the error.

Researchers at the Technical University of Dresden offered a new method for identifying mobile phones with GSM based on the physical characteristics of the radio equipment without the use of the IMSI SIM-card identifier and the identifier of a particular mobile phone IMEI [49]. There are suggested to use characteristic error patterns that are individual for each mobile device. The individuality of a mobile device is due to the presence of tolerances in the characteristics of the radio frequency path in the class of fault-tolerance due to inaccuracies in the production process and the breakdown of the parameters of components. The research has proved the practical implementation of the authentication procedure based on differences in the physical level of mobile devices that can be tracked by GSM technology.

In the process of forming the output modulated signal there are unpredictable errors that can be measured and used to create a unique template for a mobile device. The algorithm for creating a template is as follows.

(1) When receiving a radio signal in the demodulator, an estimation of the received symbol is carried out by the method of maximum likelihood. The best approximation to the transmitted symbol is chosen, that is, the nearest point of the signal constellation diagram in terms of the Euclidean metric. If the signal error is large enough, then a point other than the transmitted one can be selected. Then the demodulator produces an incorrect result.

(2) Based on the error vector for each transmitted symbol, the metric of the phase error is calculated. Research has shown that phase error is the most informative for creating a mobile device template.

(3) The trajectory of the phase error signal metric for the data packet is calculated as a percentage deviation from the mean value of the phase error, depending on the packet character transmitted. This trajectory is a template for a mobile device.

Further, with the help of neural network technology, namely machines of linear reference vectors, training is carried out, which allows you to authenticate a mobile device by template.

The estimation of the efficiency of the method was based on the TAR (True Acceptance Rate)—the probability of correct identification of the mobile device and amounted to 96.67%. At the time of publication, thirteen mobile phones were investigated, four of which belonged to the same model. It is shown that thirty packets are enough to create a test template.

The positive aspects of the study include the fact that the authentication of a mobile device does not need to crack the encrypted GSM traffic, in addition to the model of mobile device to fake is almost impossible. The principal note is that the template is not actually created by the mobile device itself, but by the pairing the mobile device is the receiver. The receiver makes its own mistakes in decoding the signal, so it's not correct to talk about a mobile device template, but a couple of patterns. The authentication of a mobile device for the purpose of digital criminology must be carried out with the help of the receiver, which created a unique template.

The need for two-factor authentication of the user in the online service to which he accesses through a personal computer is solved with the help of the second factor of the authentication—a mobile phone. The system server can check whether the computer and mobile phone are located next to each other, comparing their GPS coordinates. GPS sensors are available on all modern handsets, but rarely found on commercial computers. Researchers at the Institute for Information Security in Zurich have proposed ways to conduct two-factor authentication based on the sounds of the environment [50]. The second factor in authenticating the method is the proximity of the user's mobile phone to the device (personal computer, laptop) used to log on. The closeness of the location of two devices (mobile phone and computer) is verified by comparing the noise of the environment, which is recorded by the microphones of the devices. A significant advantage of the proposed method is user convenience—authentication takes place without any action on its part.

Since the basis of the method is the comparison of the noise of the environment of both devices, which depends on the location of the entry point (university audience, cafe, station, etc.), the noise pattern in the usual sense is not created. Instead, the similarity of the sounds simultaneously recorded by the computer and the mobile phone is compared. For this purpose, the cross-correlation function of the amplitude of the noise of the first $x(i)$ and the second $y(i)$ devices is used.

$$c_{x,y}(l) = \sum_{i=0}^{n-1} x(i) \cdot y(i - l), \qquad l \in [0, n - 1] \tag{6}$$

To avoid the effect of the difference in the recording paths of both devices, the cross-correlation function is normalized to the mean square noise deviation

$$c'_{x,y}(l) = \frac{c_{x,y}(l)}{\sqrt{c_{x,x}(0) \cdot c_{y,y}(0)}} \tag{7}$$

To calculate the similarity of noise recordings, each device uses the following algorithm for noise pollution of the environment.

(1) The noise of the environment is recorded by each device. With the help of m digital filter system, m sound records are generated for each frequency band. The number of frequency bands is 32.

(2) For m pairs of records $x(i)$ and $y(i)$, m normalized cross-correlation functions are calculated. For each of them there is the maximum value $\max(c_{x,y}^m(l))$.

(3) Similarity estimation is calculated on the basis of averaging values of maximum cross-correlation functions from the pairs of signal components

$$S_{x,y} = \frac{1}{m} \sum_{i=1}^{i=m} \max(c_{x,y}^m(l)) \tag{8}$$

Of interest is the authentication protocol, also proposed by the authors of the work. When registering a user on the service, the server provides it with an open key that is written to the mobile phone. When a user enters a login and password through a browser, the server checks the validity of the password and sends the command to start recording the environment noise on both the mobile phone and the computer, which is further forwarded by the public key. After recording, the computer encrypts the audio file with the public key of the phone and sends the cipher graph to the server, which in turn sends it to the mobile phone. The mobile phone decodes the audio file and then the calculation procedure for estimating the similarity of the noise signals takes place. Finally, the phone tells the server whether it is possible to consider either two devices next to each other. The server either allows you to log on to the system, or discards it.

The conducted experiments made it possible to determine the parameters of this method of checking the proximity of the location of both devices, the main of which is the length of the audio file, the distance between the devices and the band. A significant plus of the method is its full automatic operation in real time. Nevertheless, the authors emphasize that this method does not protect against heightened attacks. You should also pay attention to a fairly high level of hardware and software. The experiments used iPhone 5 and Nexus 4 mobile phones, MacBook Pro laptops and Dell. FRR was 0.002. The authentication time was about 5 s. You must also forward an encrypted audio file with duration of 3 s.

This method is well suited for a specific contingent of users who use similar mobile devices and computers and who do not have any restrictions on the use of devices in this class. This class of tasks requires real-time solutions in conditions of complete automation.

5   Detection of Counterfeit and Counterfeit Products

Counterfeit notebooks and mobile phones are a serious problem, especially for developing countries. The situation leads to a loss of profits from manufacturing companies, negatively affects brands, as well as environmental status. Detection of counterfeit and counterfeit goods requires cost-effective solutions. Uniform, cost-effective solution for product identification today is the paper bar codes that can easily be damaged. RFID tags are expensive; their widespread use is possible provided that the price of one label does not exceed 5 cents, while the real price is tens of times higher. Therefore, the identification of electronic devices for their individual physical characteristics is a promising effective solution to the problem of counterfeiting.

In 2012, the standard ISO 12931: 2012 (confirmed in 2017) was published, intended to assist organizations that have to verify the authenticity of material goods [51]. In 2015, a report [52] was published, which presents categories of authentication elements to reduce the risks of counterfeit products.

Counterfeit electronic devices are mostly different from genuine component parts with a lower resistance group to voltage, current, frequency, temperature, pressure, radioactive background, and so on. To detect this, they resort to a variety of tests:

- The parametric tests are a variety of electrical tests (contact test, power consumption test, output current rise time's tests, etc.). Highly qualified personnel and specialized laboratories are required;

– The functional tests are the most effective means of verifying the functionality of components that have a high degree of complexity. Is one of the most expensive testing methods to detect counterfeit devices;
– The pass tests are intended to determine the failure or abnormal operation of the component in stressed working conditions, for example, at elevated temperatures. The problem with this testing is that the device is damaged or damaged. Highly qualified personnel required;
– The structural tests reveal possible defects and anomalies associated with internal structures. In these tests, decapsulation is used—removal of the outer part of the electronic component for inspection of the inside. In this case, chemically active substances can be used, or the body of the body is broken. In any case, the device is destroyed.

All of these tests are carried out in specialized laboratories, where you want to bring the investigated device. Personnel must have high qualifications. As a result of pass and structural tests the device will be destroyed.

The same report discusses possible trials that are based on other approaches. The basic idea is that counterfeit electronic devices consist of cheap components, or they use more simple and cheap manufacturing methods. This is reflected in the characteristics of the unintentional radiofrequency radiation of the operating device, which differ significantly from similar radiation of the undisturbed device [53, 54]. There are also differences in radio frequency emissions when transmitting signals. For example, the GSM mobile phone's signal during the talk has unique features that are related to radio components (filters, amplifiers, external interfaces). The difference between the signals from two phones can be used to identify the phones, and also you can distinguish a fake phone from the real one. The accuracy of the method can be very high, from 94 to 100%, depending on the phone model [49]. Another test is based on the use of physically defunct functions (PUFs) that provide a reliable and easy-to-evaluate device authentication result [55].

New approaches provide consistently high-precision results. In order to forge the results of testing, one would have to use high-quality equipment for mobile phones and computers, which would negate the economic benefit of the production of counterfeit products.

Detection of counterfeit and counterfeit products by the method of recognition of a unique radio frequency radiation device requires only the switching of the electronic device and measurement of this radiation, which is easy to implement in any city. This eliminates the requirement for high qualifications of the operator conducting the inspection. The device does not collapse. The check takes a little time and is economically justified.

6 Authentication of Sensors

Sensors are primary measuring transducers, which, as a result of interaction with the object of measurement, produce signals and, through the interface, transmit them to the system of registration or automatic control [56]. The most common are sensors with an electrical output signal, which is generated by components—electronic components. The noise of electronic components working sensors can be used for their authentication. Such an opportunity is demonstrated in [57, 58] for sensors for measuring the radiation level of neurons and gamma radiation controlled by the international organization of the IAEA. Ensuring the authenticity of the signal from sensors controlled remotely is the basis of the guarantee of the findings of the IAEA. The National Laboratory of Idaho, USA, carried out works that proved the possibility of authentication of sensors of radiation levels of neurons (3 identical sensors) and gamma radiation (2 identical sensors) by passive noise analysis.

An electrical signal from the output of the sensor is the sum of the useful signal of the event and noise recording. In order to authenticate noise, a useful signal for recording radioactive radiation is removed by passing the output signal through the Butterworth filter. A useful signal from the passage of a neuron or a gamma-particle through a sensor is recorded as a short-term surge (~1 ms for the passage of the neuron) of the amplitude of the output signal of the sensor.

The template for each sensor is created as a result of the next processing of the noise signal:

– The amplification of the noise signal in the time domain several hundred times to get into the dynamic range of the ADC.
– Characterization of noise signal with sampling rate of 1 MHz.
– Receiving the noise spectrum using FFT, the distance between the frequency components was 30 Hz for the test stand and 60 Hz for the stand sample, made for use for identification purposes.
– Rejection of the noise, the spectral density of which is less than the given threshold. The threshold entry is logically related to the fact that the ADC has its own noises that are added to the noise being studied. The ADC noise estimation must be performed before the experiments, but they can be estimated as the value of the minimum signal level that can be measured by a specific ADC.
– Multiplication of the spectrum of noise on the transmission path mask is a measurement that is obtained without a connected sensor.
– Averaging many noise spectra to eliminate random noise.
– Storage of the spectrum of noise as a template of a specific sensor in the database of 5 sensors.

In this study, the sensor pattern has a basic spectral peak (its frequency and height) and its harmonics, as well as their width. Researchers associate such a spectrum with an impulse power supply of the sensor.

To verify the authentication of each of the sensors, an identification scheme (one to many) was used and a robust positive result of unambiguous sensor recognition was obtained. The distance between the template from the database of the verifiable sensor and the re-measured template is at least two times smaller than the distances to the templates of other similar sensors. In parallel, it was discovered that sensor patterns individually react to ambient temperatures, which could also serve as an additional feature of authentication. To do this, you need to compare the change of the sensor pattern with the temperature changes indoors.

A significant plus of the investigated authentication scheme of sensors is its remoteness. Application of the same mask of the measuring path is problematic. At first glance, the use of a mask gives the opportunity to have at the output of the algorithm of the periodogram of noise only the investigated sensor. But due to the unpredictable effects of external electromagnetic fields, the mask obtained in previous measurements may not coincide with the mask obtained with subsequent measurements. Then this can lead to nasty authentication results. In addition, when changing the configuration of the transmission path, you need to re-measure to create a mask pattern. In addition, the orientation to the identification of only the sensor from the entire measuring and transmission path may lead to not tracking other unauthorized actions aimed at falsifying the monitoring data of the radioactive situation.

Attention is drawn to the proven base of sensors, which is explained by the specifics of their use. It is likely that non-recognition of the same type of sensors could be detected due to the fairly low spectral resolution of the stand of authentication for a larger evidence base. The dependence of the temperature sensor pattern on the temperature can also result in inadequate averaging results, as the average estimates for templates may be shifted.

The main purpose of authentication of sensors is the need to obtain reliable information from them. Sensors can be installed in hard-to-reach places for regular checking, for example, places with a high radioactive background. Visual monitoring of sensors may be difficult. It should also be borne in mind that control should be carried out throughout the measurement path—data transmission. There should be a guarantee that there was no unauthorized connection to the cables, for example, generators of transmission of false signals. Therefore, we need to talk about the authentication of the measurement path—the transmission of signals.

The passive noise analysis method can be used to indicate unauthorized interference. The basic idea of detecting falsifications is based on the assumption that each element of the measurement path—the signal transmitted forms its characteristic noise into the generic signal that reaches the authentication block. To do this, set a metric to detect differences in the energy spectrum and use it to detect unauthorized interference. For greater accuracy of fixation of interventions conduct observation and control in different spectral ranges. The authentication algorithm consists in measuring the voltage at the output from the signal transmission path, applying FFT to obtain the spectrum and comparing it with the base spectrum. In the general form of the spectrum, it is possible to distinguish several different components, each of which is responsible for different phenomena. Firstly, this is the spectrum of the signal of registration of radioactive radiation, and secondly, it is smaller in height spectral stoves characterizing the sensor and signal amplifier. Thirdly, it is a component with a minimum spectral density, which is introduced by ADCs, cables, switching and connectivity devices.

Intervention experiments were performed for various attack scenarios: replacement of sensors, cable replacement, disconnection facts—cable connections, as well as connection of oscillograph and robotic signal generator cables [59]. Depending on the type of attack, there were different variations of the change in the spectral composition of the noise—an increase in the overall level of the minimum noise, as well as the shift of spectral peaks. The results of the experiments indicate that using the radio frequency spectrum of noise and impulses can detect and recognize most of the attack scenarios, including the most critical one—connecting a device with low impedance to signal falsification, disconnecting the cable, removing the sensor, replacing it initial amplifier.

It has been experimentally established that the difference in authentication attributable to unauthorized interference depends on the distance between the point of intervention and the authentication block. With increasing distance there is a leveling of signs of unauthorized interference, for which the spectral sheerness of the noise is much smaller than the spectral density of the useful signal. If you simulate the transmission path as a set of blocks, each of which is characterized by the values of attenuation of the electric signal and own noise, it is clear that with an increase in the number of such units, the transmission path "forgets" the noise that was at its inception. This fact causes the restriction on the use of this method of detecting unauthorized interference. Thus, detection of unauthorized access with remote signaling can be encountered with a number of problems.

7 Authentication of Personal Computers and Laptops

Authentication of personal computers and laptops by physical properties is important for many reasons. This may be a protection against the substitution of logical addresses for identification on the corporate network and the Internet, authentication of inventory. With this authentication, you can find out if unauthorized replacement of components has occurred. You can also monitor the quality of computer production, and much more. Therefore, the interest in the tasks of authenticating PCs and laptops is quite large. To date, the task of authenticating PCs and laptops are appropriate in terms of the study of the spectral composition of their electromagnetic radiation. Similar tasks exist for the authentication of gadgets and a wide range of electronic equipment that is used on a daily basis.

Traditionally, electromagnetic radiation of electronic devices is considered as a random system noise, the intensity of which should be below a certain threshold for compliance with state standards. Each electronic device is characterized by a unique electromagnetic radiation that can be used as a radio frequency identification device. From the spectrum for a certain frequency range, as a rule, indicates the type of device: PC, home appliances, gadgets, power tools, cars; this experiment was proved experimentally [60]. Classification of the device type was performed on the basis of the reference vector (SVM) method, which contains complex calculations and requires a learning procedure. Actually, the electromagnetic radiation of the device can be used as an identification tool, instead of using the technology of RFID tags. This can significantly reduce the cost of logistics operations for asset management and inventory tracking.

The idea of using electromagnetic radiation for the identification of laptops, smartphones, and other electronic devices has been developed in work [61]. Consequently, the technology under consideration is proposed for the identification of laptops and other devices with the help of an external registrar. The identification procedure requires direct access to the devices under study, which must be in working order.

Creating a template for electromagnetic radiation in principle is similar to the procedure for sensors, with the difference that in this case, as a signal recorder, an RTL-SDR (monopoly antenna) reader is used on the basis of the Realtek RTL2832 chip. The chip contains two 8-bit ADCs with a sampling rate of up to 3 MHz and evaluates the signal from I/Q modulation. The algorithm is the next.

(1) Receiving and evaluating the signal from the enabled device with a sampling frequency of 1 MHz. A preliminary study showed that the signals are located in the low-frequency region, which explains the choice of sampling frequency. Reading time takes a fraction of a second.
(2) The use of FFT with a countdown of 2^{17}, which gives a resolution of 7.6 Hz.
(3) Removing noises with low values of spectral density on the basis of the set *Threshold*, which is 1% higher than the difference between peak and average (*mean*) spectral depth of the noise signal: *Threshold* = ((*peak* − *mean*) 1% + *mean*).
(4) Points that appeared above the threshold are written in the form of value pairs (frequency-magnitude). A set of value pairs forms a device template. The default template contains from 1000 to 2000 points.
(5) The template is stored in the database.

For uniquely identifying a device, a 2-step ranking procedure is used, which results in the device being originally classified by type (laptop, mobile phone, etc.), and then there is an individual identification within the type.

The device type rating is by comparison of frequency sets, and individual identification within the type occurs by comparing the magnitudes corresponding to the frequencies.

The cosine similarity is used for the ranking, C.S. [62, 63]. One of the advantages of cosine similarity is the possibility of its rapid realization, especially for sparse vectors, for which only non-zero values are to be taken into account. Let $\{X_i\}$ and $\{Y_i\}$ be two sets that form two templates where X_i, Y_i are the magnitudes of the spectral components of electromagnetic radiation from the device X and device Y. Then the cosine similarity of the sets is determined by the expression:

$$C.S. = \frac{\sum_{i=1}^{N} X_i Y_i}{\left(\sum_{i=1}^{N} X_i^2 \sum_{i=1}^{N} Y_i^2\right)^{1/2}} \tag{9}$$

Different devices within the same type have greater cosine similarity than devices of different types. The advantage of using cosine similarity for ranking is that its result is invariant to different coefficients of amplification of the electromagnetic radiation registration module. Thus, the device template can be obtained with a single reader, and checked with another; in addition, the angle between the antenna plane and the surface of the device should not be strictly supported.

Identification within a single type of device using cosine similarity is based on the principle of maximum similarity. As a result of the "one-to-many" check, there is a template whose cosine similarity with the unknown device will be maximized.

Similar devices may look very similar to templates. Since, for each identification, the template for new measurements from one device is usually different from what is written in the database, the situation may occur that the device will be erroneously identified. It turned out that for different types of devices by the method of electromagnetic radiation the probability of correct identification is different. Thus, the greatest probability of correct identification is characterized by computer monitors, less likely to have laptops, and the smallest have mobile phones.

To investigate the identification errors, the distribution histograms of the Euclidean distances calculated between the patterns obtained as a result of many measurements of electromagnetic radiation from one device and two templates from the database, one of which is the template T of the investigated device, and the second template F taken from a similar device, are analyzed. The histograms are modeled as Gaussian distributions with mathematical expectations μ_T, μ_F and mean square deviations σ_T, σ_F. There are the greater the area of overlapping histograms (Gaussian distributions), the greater the probability of false identification. To estimate the probability of the correct identification of the device T as a device T, it is necessary to calculate the area:

$$P_T = \frac{1}{\sqrt{2\pi}} \int\limits_{\mu_T - 3\sigma_T}^{x_0} \left(\frac{1}{\sigma_T}\exp\left(-\frac{(x-\mu_T)^2}{2\sigma_T^2}\right) - \frac{1}{\sigma_F}\exp\left(-\frac{(x-\mu_F)^2}{2\sigma_F^2}\right)\right) dx \tag{10}$$

Since 99.73% of the Gaussian is in the region $\pm 3\sigma$ relative to μ, then in the integral the lower boundary is replaced by $-\infty$ on $\mu_T - 3\sigma_T$, this insignificantly affects the accuracy of the calculation. The point x_0 is the x-coordinate of the point of intersection of two Gaussians. Similarly, it is possible to estimate the probability of a correct non-identification of the device T as an F device:

$$P_F = \frac{1}{\sqrt{2\pi}} \int\limits_{x_0}^{\mu_F + 3\sigma_F} \left(\frac{1}{\sigma_F}\exp\left(-\frac{(x-\mu_F)^2}{2\sigma_F^2}\right) - \frac{1}{\sigma_T}\exp\left(-\frac{(x-\mu_T)^2}{2\sigma_T^2}\right) dx\right) \tag{11}$$

The common area of both Gaussian estimates of the probability sum of FRR and FAR.

Hence, histogram modeling as a Gaussian distribution can be used to predict the ability of the identification system to correctly identify devices for their electromagnetic radiation. If the distance between the Gaussian centers is greater than $3(\sigma_T + \sigma_F)$, then the two devices can be virtually unambiguously identified.

The method of electromagnetic radiation gave unambiguous identification of types of devices (laptops, LCD screens, mobile phones, fluorescent lamps and light swords were investigated). Experiments on the identification of individual devices within the same type showed the following result: the five MacBook Pro notebooks had an average identification accuracy of 94.6%; twenty LCD screens, respectively 94.7%; iPhone 6 were identified less securely—with an accuracy of 71.2%; fluorescent lamps are identified with an accuracy of 86%; light swords—100%.

The fact of precise identification within one type of device, as compared to others, indicates that the influence of external fields in this experiment was negligible. But the requirement for the absence of external electromagnetic fields during the identification can be practically unrealistic. Inaccurate identification of other types of devices, primarily laptops and LCD screens, may be explained by several reasons. The main reason most likely is the error of the RTL-SDR reader, which can give rise to a breakdown of frequencies up to 20 Hz. The second reason may be the unsteadiness of electromagnetic radiation, which the authors of the study write nothing. It should be noted that the method is intended for operating devices, whereas an RFID tag identifies the device regardless of its on/off state.

This method requires an external meter and operator, so it is not suitable for identifying computers on the network [64–69].

8 Authentication of Integrated Circuits

During the manufacturing processes of integrated circuits there are random deviations from the specified parameters: the concentration of donor or acceptor impurities, the geometry of transistors, the thickness of the oxide layer, or other technological inaccuracies. This leads to the disparity of each microscopic transistor of TTL integrated circuits (transistor-transistor logic) [70], which in turn affects the values of the threshold voltages of the transistors, the rate of increase and decrease of the fronts of pulses, the various delays in the propagation of signals, depending on the path, and so on. These uncontrolled phenomena are used to identify digital devices. For example, the manufacturers of FPGA Xilinx and Altera (Intel) use PUF as a built-in non-cloned FPGA identifier (programmable logic integral circuit) [71, 72].

Physically non-cloned functions based on signal delay (ring generators) use the difference in time of passing several copies of one signal through the given configuration of symmetric paths. In the given query in binary form, in this case there is a configuration of the paths; the answer is the result of comparing the delays of the propagation time of the signals [73].

Physically non-cloned static memory (SRAM) functions use the uniqueness of the bits values stored in each memory element. As a result of the resignation of the memory elements (field transistors) there is a change in the values (0 or 1), for each chip it is a unique configuration [74].

One of the main problems of the PUF example given on the basis of static memory is the instability of some state values, which requires the use of error correction codes. The second problem is the aging of floating gate field transistors, which leads to a change in the chip ID.

New chip authentication technology developed by Toshiba. The basis is the random RTG (Random Telegraph Noise) in transistors, which occurs due to defects in the insulation material [75]. This kind of PUF based on RTN is more stable than previous electrical loads, actual measurement data can be used more than one million times. Toshiba is developing inter-authentication technology for IoT devices by removing PUF reflections in semiconductor chips for fastest deployment to create safer IoT systems.

Toshiba does not disclose the details of the use of RTN for authentication, but based on the general characteristics of this type of noise [76, 77]—the concentration of spectral density in the low frequency region (≈ 1 Hz)—it can be refined that the authentication procedure will be sufficiently long imposed Limiting the speed of electronic devices.

9 Conclusions

Receiving, processing, transferring and storing business information is carried out with the help of various electronic devices that work as part of information and communication systems, or are supplied to the market for these purposes. At the same time, the need to authenticate electronic devices, in particular, to counteract cyberattacks, the distribution of counterfeit products and media content, is increasing to enhance the security of financial transactions and access to corporate or state online services. The wide range of electronic devices, their functional capabilities and applications in business informatics determines the relevance of the classification of authentication tasks and the formation of a general approach to the authentication of electronic devices.

The tasks that are solved on the basis of authentication are proposed to be divided into six classes, and as a general approach, use the authentication of electronic devices based on their individual differences. The latter is due to the fact that the production of electronic devices ensures their operation within the fault tolerance and allows for some physical non-identity.

References

1. Kryvinska N (2012) Building consistent formal specification for the service enterprise agility foundation. J Serv Sci Res 4(2):235–269 (Springer, The Society of Service Science)
2. Kaczor S, Kryvinska N (2013) It is all about services—fundamentals, drivers, and business models. J Serv Sci Res 5(2):125–154 (Springer, The Society of Service Science)
3. Gregus M, Kryvinska N (2015) Service orientation of enterprises—aspects, dimensions. Comenius University in Bratislava, Technologies
4. Kryvinska N, Gregus M (2014) SOA and its business value in requirements, features. Comenius University in Bratislava, Practices and Methodologies
5. Molnar E, Molnar R, Kryvinska N, Gregus M (2014) Web Intelligence in practice. J Serv Sci Res 6(1):149–172 (Springer, The Society of Service Science)
6. Kirichenko L, Radivilova T, Zinkevich I (2017) Forecasting weakly correlated time series in tasks of electronic commerce. In: Proceedings of 12th international scientific and technical conference on computer sciences and information technologies (CSIT) 2017, Lviv, Ukraine, pp 309–312. https://doi.org/10.1109/STC-CSIT.2017.8098793
7. Kirichenko L, Radivilova T, Zinkevich I (2018) Comparative analysis of conversion series forecasting in E-commerce tasks. In: Shakhovska N, Stepashko V (eds) Advances in intelligent systems and computing II. CSIT 2017. Advances in intelligent systems and computing, vol 689. Springer, Cham, pp 230–242. https://doi.org/10.1007/978-3-319-70581-1_16
8. Dobrynin I, Radivilova T, Maltseva N, Ageyev D (2018) Use of approaches to the methodology of factor analysis of information risks for the quantitative assessment of information risks based on the formation of cause-and-effect links. In: Proceedings of the 2018 international scientific-practical conference problems of infocommunications. science and technology (PIC S&T), Kharkiv, Ukraine, pp 229–232. https://doi.org/10.1109/INFOCOMMST.2018.8632022
9. Kirichenko L, Bulakh V, Radivilova T (2017) Fractal time series analysis of social network activities. In: Proceedings of the 4th international scientific-practical conference problems of infocommunications. Science and technology (PIC S&T), Kharkov, Ukraine, pp 456–459. https://doi.org/10.1109/INFOCOMMST.2017.8246438
10. Kirichenko L, Radivilova T, Tkachenko A (2019) Comparative analysis of noisy time series clustering. In: Proceedings of the 3rd international conference on computational linguistics and intelligent systems (COLINS-2019), 18–19 April, Kharkiv, Ukraine, vol 1, pp 184–196
11. Ageyev D (2010) NGN network planning according to criterion of provider's maximum profit. In: Proceeding of the IEEE international conference on modern problems of radio engineering, telecommunications and computer science (TCSET-2010), 23–27 February, Lviv, Ukraine, pp 256–256
12. Ageyev D, Al-Ansari A, Qasim N (2015) Multi-period LTE RAN and services planning for operator profit maximization. In: Proceeding of the IEEE the experience of designing and application of CAD systems in microelectronics, 24–27 February, Lviv, Ukraine, pp 25–27. https://doi.org/10.1109/CADSM.2015.7230786
13. Ageyev D, Wehbe F (2013) Parametric synthesis of enterprise infocommunication systems using a multi-layer graph model. In: Proceeding of the IEEE 23rd international crimean conference microwave and telecommunication technology, 8–14 September, Sevastopol, Ukraine pp 507–508
14. Al-Dulaimi A, Al-Dulaimi M, Ageyev D (2016) Realization of resource blocks allocation in LTE downlink in the form of nonlinear optimization. In: Proceeding of the IEEE 13th international conference on modern problems of radio engineering, telecommunications and computer science (TCSET), 23–26 February, Lviv, Ukraine, pp 646–648. https://doi.org/10.1109/TCSET.2016.7452140
15. We Are Social Homepage (2019). Available at: https://digitalreport.wearesocial.com/
16. International Organization for Standardization (2005) ISO/IEC 19794-2:2005 Information technology—biometric data interchange formats—part 2: finger minutiae data. Available at: https://www.iso.org/standard/38746.html

17. International Organization for Standardization (2017) ISO/IEC 19794-15:2017 information technology—the format of the exchange of biometric data—part 15: image of wrinkles in the palm of your hand. Available at: https://www.iso.org/standard/63865.html
18. Khana SH, Akbar MA, Shahzad F, Farooq M, Khan Z (2015) Secure biometric template generation for multi-factor authentication. Pattern Recogn 48(2):458–472
19. Pankanti S, Bolle RM, Jain AK (2000) Biometrics: the future of identification. IEEE Comput 33(2):46–49
20. Kim JO (1989) Factor, discriminant and cluster analysis. Finance and Statistics, Moscow
21. One-to-Many Identification (2009) In: Stan ZL, Anil KJ (eds) Encyclopedia of biometrics. Springer, Boston, MA
22. Identification (2009) In: Stan ZL, Anil KJ (eds) Encyclopedia of biometrics. Springer, Boston, MA
23. Open-Set Identification (2009) In: Stan ZL, Anil KJ (eds) Encyclopedia of biometrics. Springer, Boston, MA
24. Chaplyga V, Nyemkova E, Ministr J, Chaplyga V (2018) Fast algorithms for deterministic non-equidistant digital filtering of signals in the time domain. In: Proceedings of the 2018 international scientific-practical conference problems of infocommunications. Science and technology (PIC S&T), Kharkiv, Ukraine, pp 135–139. https://doi.org/10.1109/INFOCOMMST.2018.8632145
25. Chaplyha V, Nyemkova E (2017) Using non-uniform sampling in real-time correlation processing of authentication signals. In: Proceedings of the 2017 international scientific-practical conference problems of infocommunications. Science and technology (PIC S&T), Kharkiv, Ukraine, pp 474–476. https://doi.org/10.1109/INFOCOMMST.2017.8246441
26. Nyemkova E, Shandra Z (2016) Modelling of the phase portraits of noise for identification of electronic devices. In: Proceedings of the 2016 international scientific-practical conference problems of infocommunications. Science and technology (PIC S&T), Kharkiv, Ukraine, pp 125–127. https://doi.org/10.1109/INFOCOMMST.2016.7905356
27. Rybalsky OV, Soloviev VI, Shablya AN, Zhuravel VV, Solovyov VI (2015) System of automated voice search. Inf Math Methods Model 5(4):302–307
28. Rybalsky OV, Zhuravel VV (2018) Method of verification of a person according to physical parameters of speech-free signals of messages contained in a voice database, using the automatic search system "AVATAR". State Research Experimental Forensic Center of the Ministry of Internal Affairs of Ukraine, Kyiv
29. Rybalsky OV, Soloviev VI, Zhuravel VV (2016) Methodology for constructing a system of expert verification of digital phonograms and identification of digital recording equipment with the use of the fractal program. Inf Math Methods Model 6(2):105–115
30. Rybalsky OV, Zhuravel VV, Soloviev VI, Zheleznyak VK (2016) A generalized model for the separation of fractal structures from digital signals by the method of maximums of wavelet transformation. Bull Polytech State Univ Ser S Fundam Sci 4:13–16
31. Zhuravel VV (2016) Principles used in designing software to verify the integrity of information in digital phonograms. Mod Spec Technol 2(45):39–43
32. Rybalsky OV, Solovie VI, Zhuravel VV, Shablia AN, Tatarnikova TA (2014) Automatic calculation of the coefficients of the fractal scale in the program "Fractal". Mod Spec Techn 4(39):5–11
33. Rybalsky OV, Zhuravel VV, Soloviev VI, Timoshenko LN (2016) Statistical processing of self-similar structures isolated from phonogram sounds, when determining the identity of digital sound recording equipment. Electr Comput Syst 22(98):413–417
34. Rybalsky OV, Zhuravel VV, Soloviev VI (2016) Determination of the effectiveness of the tools of examination of materials and means of digital sound recording "Fractal". Topical Issues of Expert Crimean Law Enforcement of Law Enforcement Activities, National Academy of Internal Affairs, Kyiv, pp 251–253
35. Rybalsky OV, Soloviev VI, Zhuravel VV (2018) The systems of tool of examination of audio and video recordings are in Ukraine. Bull Polotsk State Univ Ser C, Fundam Sci 4:1–15

36. Rublev DP, Chumachenko AV, Makarevich OB, Fedorov VM (2007) Identification of digital microphones under the needelality of recording tract. Bull Southern Federal Univ Tech Sci Inf Secur 1(76):84–92
37. Hua G, Bi G, Thing VLL (2017) On practical issues of electric network frequency based audio forensics. IEEE Access 5:20640–20651
38. Maher RC (2009) Audio forensic examination: authenticity, enhancement, and interpretation. IEEE Signal Process Mag 2(2):84–94
39. Gerazov B, Kokolanski Z, Arsov G, Dimcev V (2012) Tracking of electrical network frequency for the purpose of forensic audio authentication. 13th international conference on optimization of electrical and electronic equipment on proceedings, 2012. EPSMA, Brasov, Romania, pp 1164–1169
40. Liu Y, Yuan Z, Markham PN, Conners RW, Liu Y (2011) Wide-area frequency as a criterion for digital audio recording authentication. IEEE Power and Energy Society General Meeting, San Diego, CA, USA, pp 1–7
41. Chai J, Liu F, Yuan Z, Conners RW, Liu Y (2013) Source of ENF in battery-powered digital recordings. Audio Eng Soc Convention 135:1–6
42. Liu Y, Yuan Z, Markham PN, Conners RW, Liu Y (2012) Application of power system frequency for digital audio authentication. IEEE Trans Power Delivery 27(4):1820–1828
43. Lv Z, Hu Y, Li C-T, Liu B-B (2013) Audio forensic authentication based on MOCC between ENF and reference signals. IEEE China summit and international conference on signal and information processing, 2013. Beijing, China, pp 427–431
44. Chuang W-H, Garg R, Wu M (2013) Anti-forensics and countermeasures of electrical network frequency analysis. IEEE Trans Inf Forensics Secur 8(12):2073–2086
45. Hu G, Zhang Y, Goh J, Thing VLL (2016) Audio authentication by exploring the absolute-error-map of ENF signals. IEEE Trans Inf Forensics Secur 11(5):1003–1016
46. Elmesalawy MM, Eissa MM (2014) New forensic ENF reference database for media recording authentication based on harmony search technique using GIS and wide area frequency measurements. IEEE Trans Inf Forensics Secur 9(4):633–644
47. Karantaidis GD (2017) Multimedia authentication using electric network frequency. Department of Informatics, Faculty of Sciences Aristotle University of Thessaloniki, Thessaloniki
48. IEEE: Standard for Information technology (2016) IEEE 802.11-2016—Telecommunications and information exchange between systems local and metropolitan area networks-specific requirements—part 11: wireless LAN Medium Access Control (MAC) and Physical Layer (PHY) specifications. Homepage, Available at: https://standards.ieee.org/standard/802_11-2016.html
49. Hasse J, Gloe T, Beck M (2013) Forensic identification of GSM mobile phones. In: Proceedings of the first ACM workshop on Information hiding and multimedia security, IH&MMSec'13, 2013, New York, NY ACM, Montpellier, France, pp 131–140
50. Karapanos N, Marforio C, Soriente C, Čapkun S (2015) Sound-proof: usable two-factor authentication based on ambient sound. In: Proceedings of the 24th USENIX security symposium. USENIX Association, Washington D.C., pp 483–498
51. International Organization for Standardization (2012): ISO 12931:2012—performance criteria for authentication solutions used to combat counterfeiting of material goods. ISO Homepage, Available at: https://www.iso.org/standard/52210.html
52. Baldini G, Fovino IN, Satta R, Tsois A, Checchi E (2016) Survey of techniques for fight against counterfeit goods and Intellectual Property Rights (IPR) infringing. Publications Office of the European Union, Belgium, Brussels
53. Cobb WE, Laspe ED, Baldwin RO, Temple MA, Kim YC (2012) Intrinsic physical-layer authentication of integrated circuits. IEEE Trans Inf Forensics Secur 7(1):14–24
54. Williams MD, Temple MA, Reising DR (2010) Augmenting bit-level network security using physical layer RF-DNA fingerprinting. In: Proceedings of the IEEE global telecommunications conference GLOBECOM, 2010, Miami, FL, USA, pp 6–10

55. Tzenbeisser S, Ünal K, Rožic V, Sadeghi A-R, Verbauwhede I, Wachsmann C (2012) PUFs: myth, fact or busted? A security evaluation of Physically Unclonable Functions (PUFs) cast in silicon. In: Prouff E, Schaumont P (eds) 14th international workshop cryptographic hardware and embedded systems—CHES 2012. Lecture notes in computer science, vol 7428. Springer, Heidelberg, pp 283–301

56. METROLOGY Homepage (2019) Available at: http://metrology.com.ua/download/dstu-gost-gost-r/60-dstu/97-dstu-2681-94

57. Svoboda J, Schanfein M (2012) Transducer signal noise analysis for sensor authentication. In: PREPRINT INL/CON-12-24524, Idaho National Laboratory, Idaho

58. Svoboda J, Schanfein M (2013) Apparatus, system, and method for sensor authentication. In: JUSTIA Patents, No 10066962

59. Baker B, Sanders J, Schanfein M, Svoboda J, West J (2015) Passive noise analysis studies on tampering indication. In: PREPRINT INL/CON-15-34213, Idaho National Laboratory, Idaho

60. Laput G, Yang C, Xiao R, Sample A, Harrison C (2015) Em-sense: touch recognition of uninstrumented, electrical and electromechanical objects. In: Proceedings of the 28th annual ACM symposium on user interface software technology, ACM, NY, pp 157–166

61. Yang C, Sample AP (2016) EM-ID: tag-less identification of electrical devices via electromagnetic emissions. In: Proceedings of the 2016 IEEE international conference on RFID, Orlando, FL, USA, pp 1–8

62. Similarit, C, Bayardo RJ, Ma Y, Srikant R (2007) Scaling up all pairs similarity search. In: Proceedings of the 16th international conference on World Wide Web, ACM, Banff, Alberta, Canada, pp 131–140

63. Tata S, Patel JM (2007) Estimating the selectivity of tf-idf based cosine similarity predicates. Newsletter ACM SIGMOD Rec 36(2):7–12

64. Zhao N, Dublon G, Gillian N, Dementyev A, Paradiso J (2015) Emi spy: harnessing electromagnetic interference for low-cost, rapid prototyping of proxemic interaction. Proceedings on the IEEE 12th international conference on wearable and implantable Body Sensor Networks (BSN). MA, USA, Cambridge, pp 1–6

65. Cass S (2013) A 40 dollar software-defined radio. IEEE Spectr 50(7):22–23

66. Penney GP, Weese J, Little J, Desmedt P et al (1998) A comparison of similarity measures for use in 2-d-3-d medical image registration. IEEE Trans Med Imaging 17(4):586–595

67. Mihalcea R, Corley C, Strapparava C (2006) Corpus-based and knowledge-based measures of text semantic similarity. In: Proceedings of the 21st national conference on Artificial intelligence AAAI'06, vol 1. AAAI Press, Boston, Massachusetts, pp 775–780

68. Huang A (2008) Similarity measures for text document clustering. In: Proceedings of the sixth New Zealand computer science research student conference, NZCSRSC, Christchurch, New Zealand, pp 49–56

69. Holcomb DE, Burleson WP, Fu K (2009) Power-up sram state as an identifying finger-print and source of true random numbers. IEEE Trans Comput 58(9):1198–1210

70. Böhm C, Hofer M (2012) Physical unclonable functions in theory and practice. Springer, New York, NY

71. XILINX Security Working Group Homepage (2016) Available at: https://www.xilinx.com/about/security-working-group-2016.html

72. INTEL Homepage (2019) Available at: https://www.intel.com/content/www/us/en/programmable/support/quality-and-reliability.html

73. Suh GE, Devadas S (2007) Physical unclonable functions for device authentication and secret key generation. In: Proceedings of the 44th ACM/IEEE design automation conference, San Diego, California, USA pp 9–14

74. Holcomb DE, Burleson WP, Fu K (2008) Power-up SRAM state as an identifying fingerprint and source of true random numbers. IEEE Trans Comput 58(9):1198–1210

75. TOSHIBA Corporation Homepage (2019) Available at: http://www.toshiba.co.jp/rdc/rd/detail_e/e1806_02.html
76. Kirton MJ, Uren MJ, Collins S, Schulz M, Karmann A, Scheffer K (1989) Individual defects at the Si: SiO_2 interface. Semicond Sci Technol 4(12):1116–1126
77. Realov S, Shepard K (2013) Analysis of random telegraph noise in 45-nm CMOS using on-chip characterization system. IEEE Trans Electron Devices 60(5):1716–1722